Chintz and Cotton

India's Textile Gift to the World

Joyce Burnard

Kangaroo Press

*To the memory of my husband Eric,
and my friend Robin*

Acknowledgments

Many people gave me generous help and support in the writing of this book. Special thanks are due to Dr Lotika Varadarajan, Indian textile scholar of New Delhi, who sent me many of her own published articles and kindly answered my questions. The following read the manuscript and offered valuable criticism: Marjorie Jacobs, Emeritus Professor of History, University of Sydney; Rosalind Rennie, senior textile teacher and examiner; and Christina Sumner, Curator of Decorative Arts, Powerhouse Museum, Sydney.

Friends in India who sent valuable information were: John Bissell, Prince Jayasinhji Jhala, Lionel Paul, Kamal Singh and Neela Sontakay. In Australia generous assistance was given by Philippa Sandall, Barbara Beckett, Janet De Boer, Keith Dyson-Smith, Wilma Garnett, Moira Kerr, Liz Williamson and Kylie Winkworth. Gillian Osborne and Mona Young made important contributions by putting my typewritten copy onto disk.

I must acknowledge the assistance of the staff and the treasures of information I obtained from the following libraries: British Library (India Office), London; Calico Museum, Ahmedabad, India; Nederlands Openluchtmuseum, Netherlands; Powerhouse Museum, Sydney; State Library of New South Wales and Mitchell Library, Sydney; Textile Museum, Washington, DC; and Victoria & Albert Museum, London.

Lastly, I wish to pay tribute to two special people. One is Robin Duffecy, who loved India and introduced me to its fascinating textile side. She suggested that I write this book, and fortunately was able to read the manuscript before her untimely death in 1992. The other is my husband Eric Burnard, who died in 1991. He was my patient companion on the textile travels, and at home kept urging me to get on with writing. He would have been very pleased to see this book published.

© Joyce Burnard 1994

*First published in 1994 by Kangaroo Press Pty Ltd
3 Whitehall Road Kenthurst NSW 2156 Australia
P.O. Box 6125 Dural Delivery Centre NSW 2158
Typeset by G.T. Setters Pty Limited
Printed in Hong Kong through Colorcraft Ltd*

ISBN 0 86417 597 3

Contents

Acknowledgments 2

Foreword 4

1. Introduction – The Indian Connection 5
2. The Cotton Gift 7
3. Chintz – So English 13
4. All the Colours of the Rainbow 22
5. Dyes and Methods – The Ancient Secrets 26
6. Design: Paisley, Ikat (tie-dye), Textures 32
7. Silk in India 44
8. Mughuls and Magical Muslins 48
9. Company and Crown – Into the Raj Years 53
10. Gandhi and Threads of Independence 62
11. Revival of the Handloom 68
12. The Australian Connection 72
13. Epilogue – India Today 79

Appendix 1: Fabrics for Restoration 81

Appendix 2: Glossary 84

Bibliography 87

Index 90

Foreword

Few people, other than textile scholars, realise how much we owe to India for so many of our fabrics, especially the everyday cottons for clothing and furnishing which we tend to take for granted. And yet, of all the textile traditions of the world, the bright cottons of India have had the most profound and enduring influence on textile production worldwide.

The story which Joyce Burnard tells of how all this came about is fascinating. Readers who are not aware of the history of the association with India will be surprised at many of the author's revelations. They will discover, for instance, that names such as chintz, gingham, seersucker and calico, which we think of as so thoroughly western, all derive from Indian words because the fabric types came from India. Even familiar denim, as the author points out, originated there.

Burnard's research has been thorough, and her work is scholarly yet written in a clear style that will appeal as much to the general reader as to the textile student or specialist. She has produced both a good yarn and a very useful history of the making and merchandising of Indian textiles. This is a rich history, as colourful as the fabrics themselves, and extending back in time some 4000 years to the earliest makers and entrepreneurs. Herself an importer of handloomed Indian cottons into Australia in recent years, Joyce Burnard unwittingly joined this long line of traders and is thus an integral part of the compelling story she relates.

Christina Sumner
Curator Decorative Arts
Powerhouse Museum, Sydney

Cutwork lace, jamdani, *woven by supplementary weft technique, was exported to England from 1670. It is popular today for interior decorating. (Photographer: Andrew Payne)*

1. Introduction – The Indian Connection

What could a simple gingham dress worn to the supermarket or a seersucker table cloth used at a picnic possibly have to do with exotic Indian textiles such as shimmering saris, brocades of the princely caste, painted temple hangings, or colourful muslin turbans? As it happens they have a great deal in common. Those fabric names 'gingham' and 'seersucker' show the link. They are both derived from Indian words and there are many more like them that have become so much part of our language that we take them for granted. In fact, few people realise that our own everyday fabrics, especially our cotton ones, owe most of their origin to India...and that is what this book is all about.

I certainly had no idea of this 'Indian Connection' until I visited India for the first time with the specific purpose of buying fabrics. It was not long before my curiosity about Indian textile origins and the fascinating link with our modern fabrics was so aroused that I began on a long path of research that was to reveal some astonishing facts. It all began when a friend sent me some samples of beautiful hand-loomed furnishing cottons in a variety of textures in naturals and some colours. I loved them immediately, and when I showed them to a few interior decorating friends and their reaction was the same, I decided to import a selection. This meant visiting India, so I set off.

The friend who sent the samples met me at Delhi airport when I arrived in the middle of the night. As we drove out into the hot velvety darkness I felt instinctively that I was embarking on a significant adventure. Next morning I awoke to the bright strong light of an Indian day and, opening the window on to my host's lush garden, brought in a rush of heat that was already overwhelming at 8 a.m. It was August and not the best time of the year to visit Delhi. Seasoned residents, if they can afford the time and the money, take off to a handy hill resort for the worst of the hot weather which tails off towards the end of October.

In the days to follow I was to meet many people involved in textiles. I was directed to small handloom factories with several looms and weavers and dyers working under supervision; and to outlying villages where free-lance weavers have their looms set up in their houses or in open spaces outside, with dyers working nearby as has been done for centuries. On a more sophisticated level I met a young Sikh woman running her own block-printing factory on modern lines, her wares in demand in Paris and New York; and a general's daughter with a silk shop in New Delhi. We visited sari shops and sari makers working in narrow corners, and in Old Delhi explored the crowded bazaar where there are colourful cloth stalls everywhere.

Until I went to India I had no idea how much an integral part of life are spinning, weaving and dyeing in the production of textiles. The reasons for this are obvious; the gigantic population needs to be clothed, so there is a huge domestic market for woven cloth throughout all the great cities and in the rural areas. There is also a large export market for both hand-loomed and mill-woven products. Modern India is among the biggest producers of textiles in the world today. Watching the weavers and dyers in the hand-loom sector working deftly with confidence, I realised that they were practising ancient crafts that they had not come to lately, for their skills were born and bred. We met one weaver who told us his father had taught him weaving, and that he was the seventh generation weaver in his family. My curiosity was aroused; I wanted to know more about the traditional textiles, cotton and silk, particularly cotton which was obviously the most important and abundant yarn in India. Like many people I had thought that Egypt had been the home of cotton since ancient times, but this is not so — they had linen (flax). Cotton I was to discover quite possibly originated in India and had been spun and woven into cloth from prehistoric times. Silk also has an ancient tradition in India with its raw silk type, as distinct from the Chinese mulberry silk.

Treasure Hunt

Tracing the origins of cotton and silk and the history of Indian textiles generally and their connection with the West took me on a real treasure hunt in textile libraries and museums around the world, and back to India several times. My travels were helped by my import business which boomed in an unexpected way. Everyone who saw the beautiful Indian textured handloomed cottons loved them and wanted them. I was soon ordering huge quantities and organising their handling and distribution. Almost by chance I had founded a business which in a few years was to grow into a nation-wide concern. The business took me on regular visits to India to see suppliers, sort out problems, look for new fabrics. Unknown to me at the time I had joined a long line of traders that stretched back into prehistory, traders who had taken Indian cottons and silks and other commodities to the countries of the known civilised world. In due course the wonderful Indian textile story unfolded. What I discovered taught me much about the social history not only of India but of the world because textiles and clothing are such an essential part of daily civilised living. The parallels between the lives of ordinary working people in ancient India and our own modern world intrigued me. I found that weavers belonged to powerful guilds, rather like our unions, paid taxes and interest on loans. I also discovered that Indian colours and designs have long influenced textiles all over the world, and go on influencing them; that patterns like Paisley and Ikat and others that are perennially popular and universal came originally from India and nearby. Further, I discovered that almost every process of textile manufacture in our big modern mills such as spinning, weaving and dyeing could trace its origin to India. In short, what I discovered was the enormous textile debt which the whole world owes to India.

2. The Cotton Gift

When cotton cloth first arrived in England from India in the mid-1600s, brought by the East India Company, no one was to guess that it would change the economy not only of Britain but of the whole Western world. Who could tell then that it would play a big role in the Industrial Revolution, contribute greatly to the American Civil War, to social degradation such as child labour in England and slaves in America, but ultimately bring about radical social reform? How was anyone to know that it would become such a powerful force as to earn the name King Cotton?

India has given the world many gifts but cotton is one of the greatest, yet we take it for granted. Indeed it would be hard to imagine being without it. In every wardrobe will be found garments of cotton or cotton mixture, and in every room of the house will be cotton articles — sheets in the bedroom, towels in the bathroom, curtains and covers everywhere, cleaning cloths and mops in the kitchen. Cotton is the most practical of all yarns and is used in every country. Even its generic botanical name sounds universal — *Gossypium*.

It does seem amazing then that until little more than 300 years ago cotton cloth was virtually unknown in the West. In England the majority of people wore woollen clothes, wool being one of England's main products, or they wore fustian, a wool/linen mixture; the rich wore velvet or silk, or expensive finely woven linen from Belgium and Holland, and it was the same in France. On the other hand in almost every country throughout the Middle East and parts of the Mediterranean, and east as far as China, cotton had been in use for clothing and furnishing for more than a thousand years. In India, where it was indigenous, cotton had been grown, spun, woven, and dyed for at least 4000 years and probably much longer. The Indian people had been wearing cool cotton clothes for all that time, and had been exporting cotton to their closest neighbours.

Where did it all begin?

Early in my researches I read this sentence in a book, *Handicrafts of India* by the noted Indian woman textile historian Kamaladevi Chattopadhyay: 'The earliest madder-dyed cotton was found at Mohenjo-Daro dating it to 5000 years ago.' This presented me with a challenge. I did not know then where Mohenjo-Daro was, nor did I know about madder-dyeing, but I was determined to find out and also to learn how a piece of dyed cotton could have survived for so long and in what condition. Was the colour still in the cotton? The sentence started me off, at the beginning of the 'thread', and I have tried to follow it through to the present day.

I was to find that Mohenjo-Daro lies in Sind, a region in the Indus Valley in the far northwest of India, near Karachi, in what is now Pakistan. It is the site of two adjacent cities of the Harappa (pre-Aryan) civilisation dating from around 3000 B.C. The site was discovered in 1921 and soon afterwards a team of government archaeologists, under the direction of Sir John Marshall, Director-General of Archaeology in India, began excavations. To their amazement they uncovered evidence of a high degree of civilisation with private houses built of well-burnt bricks, public buildings including a large bath or reservoir, proper drainage systems, and well-planned wide streets. It is astonishing to realise that these cities flourished long before the first Greek cities were built.

In among the rubble was found a silver vase to the lid of

which clung a small fragment of woven cotton. It was a remarkable find because until then, due to the damp conditions in India, no samples of cotton had been found from before the 16th century. Its survival was due to a lucky accident: the formation of silver salts from the vase which had a preserving effect. Laboratory analysis revealed the cotton to be of a type similar to present-day Indian cotton and definitely not wild cotton; and even though the fragment was extremely small, experts were able to assess the density of the weaving and the weight per square yard showing that it was an open weave perhaps like muslin.[1] Some gold jewellery was found near the vase suggesting that the owner had wrapped the valuables in a whole piece of cloth and hidden them, perhaps fleeing from an invader or a natural disaster and hoping to come back later. It was thought that the cities had existed for about 1000 years until their destruction, so the civilisation must have become well-established. Unfortunately, the cotton fragment, having survived for so long, does not seem to have been kept after analysis, nor even photographed properly. The laboratory report states that it was 'very much tendered and penetrated by fungal hyphae'. It probably disintegrated with handling. A sketch of the vase appears in the published archaeological report (see illustration). Some pieces of cotton string and cord were also found attached to earthenware pots and caked in dirt. The experts could not type this cotton precisely but thought it was probably cheaper wild cotton similar to kapok. One piece of string had traces of purple dye. This was a sensational find, that the dye should still be in the fibre. Tests showed that the dyestuff was probably madder and it was concluded that as the dye had stayed in the cotton for so long a mordant must have been used. It proved that the Indians had a dye chemistry and understood fast-dyeing. Dyers' vats were found among the ruins. Spindles and spindle whorls were also found in the ruins of the houses. Spinning seems to have been a common occupation not restricted to a particular class as whorls of both expensive porcelain and cheaper pottery were found.

The Great Bath at Mohenjo-Daro, city of the Harappan civilisation c. 3500 B.C., uncovered during archaeological diggings in 1921. (Courtesy Archaeological Survey of India)

early travellers and writers. One was the Greek historian Herodotus who, in his famous history of the Greek war with Persia which took place from 490 to 480 B.C., mentioned that the Indian mercenaries fighting for the Persian King Xerxes were dressed in cotton, 'an exotic cloth that had been woven by Indian craftsmen for more than 1,000 years.' Another Greek who actually saw cotton growing in India but confused it with flax, was Nearchus, the admiral whom Alexander the Great sent to navigate the Indus River in 327 B.C. before settling a colony of Greeks in the area. Nearchus wrote: 'the Indians wore linen [sic] garments, the substance whereof they were made growing upon trees; it is indeed flax, or rather something much whiter and finer than flax. They wear shirts of the same which reach down to the middle of their legs; and veils which cover their head and a great part of their shoulders.' He was describing clothes that have not changed much over the years. It was after the arrival of the Greeks that the cotton plant and cloth began to find their way to Greece and further west. Closer neighbours, such as Mesopotamia, had been importing cotton from India for at least 2000 years before this. This was proved by Indian merchant seals being found near the site of Babylon which bore marks showing that the main commodity traded had been cotton. The Babylonians called cotton *sindhu* which seems to indicate that it came from Sind.

Greek Visitors

All this gave concrete evidence of India's ancient cotton industry. Prior to that the only knowledge had come from mentions in ancient Sanskrit literature, and from reports of

Arab Traders

In those ancient days no restrictions were placed on Indian traders travelling abroad using land and sea routes. They were excellent navigators sailing square-rigged two-masted ships across the Arabian Sea and Indian Ocean, calling at East African ports. They sailed to Indonesia where many stayed to set up trading posts and introduce Hindu culture. Trade was good in Indonesia because the people had a passion for Indian textiles which they took in exchange for spices. Many centuries later, in the Gupta period, bans were placed on Hindus leaving India. Paradoxically, this period, 320–415 A.D., also heralded a golden age for the country; trading had made it extremely prosperous and there was a great flowering of culture and ideas. At the same time the caste system was beginning to tighten, with the rise to power of the Brahmans who were now declaring it a sin for Hindus to travel abroad 'across the black waters' for fear of contamination by the impure. Indian traders had to give way to Arabs, who from the 7th century A.D. became the main traders in the whole area for hundreds of years, until the Portuguese and Dutch came on the scene in the 16th century. The Arab traders were to leave a strong mark around the region which survives to this day with Arab settlements still discernible on India's Malabar Coast, and the Muslim culture in Indonesia. In the latter country they exchanged Indian textiles for spices which they sold as far away as the Mediterranean, including Rome.

Rome was a big customer even by today's standards. At that time for wealth and power Rome would have been akin to America in the world now. The Romans placed great value on Indian cottons and silks, considering them to be of the highest quality as regards dyes and fineness of weave. It is known that Caesar's soldiers wore Indian cotton uniforms, that the Romans had tents and awnings of cotton and that they bought colourful patterned cottons from India. They charged duty on Indian textiles. They also bought spices and luxury goods such as precious stones, exotic animals and birds such as apes, parrots and peacocks. The Roman historian Pliny the Elder wrote that 'there was no year in which India did not drain the Roman Empire of a hundred million sesterces' (a sesterce would be like our dollar).

There was an enormous land trade of goods of all types up through the mountains of Afghanistan, to Persia and far beyond, and east along the Old Silk road to China, meeting up with Chinese caravans coming the other way at trading posts along the road. Cotton cloth was found among textiles, mostly silk, in tombs at Turfan in the Gobi Desert during excavations by Sir Aurel Stein early this century, and cotton which may date from then is still grown in places along the old route.

We can thank the Arabs for the name 'cotton' it comes from their word *qutan* or *kotn*. The Latin is *cotonum*; the Romans also called it *carbasina* from the Sanskrit *karpasi*. The modern Indian word is *kapas* from the same source.

All this gives conclusive proof that cotton has been growing in India and woven into cloth for at least 5000 years. Coincidentally, on the opposite side of the world, in the Americas, cotton was also growing in ancient times. The 'old' cotton belt extended from the southwest of what is now the United States, from Utah through Arizona, New Mexico and Central America, to Peru where in Pre-Columbian civilisations cotton cultivation, spinning and weaving were developed. In dry tombs of desert areas of Peru archaeologists have found cotton cloth dating back to c. 2000 B.C. The earliest fragments show weaving that is primitive but cloths from the late Inca civilisation are sophisticated and woven and dyed in a similar way to Indian cotton, which has had textile scholars wondering if there may be some connection. Designs are different, geometrical rather than the more naturalistic Indian designs, but this could be due to general differences in technique, such as Peruvian patterning in the loom as opposed to the Indian use of surface decoration, as well as cultural and climatic differences.

Ancient Cotton Links

The types of cotton growing on both sides of the world have been found to be genetically different.[2]

According to agricultural historians, the ancestry of the cotton plant goes back at least 50 million years and it originated in Africa. Over time seeds could have drifted to India and adjoining areas conducive to cotton-growing and across the sea to Central America where, over such a long period, the plants developed differently. The most used modern cottons, *G. barbadense* and *G. hirsutem*, evolved through hybridisation between the wild African species and American types.

As far as cultivation and coincidental craft skills are concerned, anthropologists think that there could be a pre-historical link. One theory is that civilisation began in Central or East Asia, where skeletal remains of Paleolithic man have been found, with people migrating not earlier than 20,000 years ago eastwards across Asia, then crossing land bridges to what is now Alaska and continuing down to warmer climates. It is thought they could have taken with them some primitive cultural knowledge that led to later development of agricultural and craft skills. Certainly the later Peruvians were known to have highly developed agricultural skills which would have enabled them to practise selective breeding of cotton.[3]

What did the cotton plant look like to early visitors to India? They would have seen a rather attractive shrub or tree with flowers vaguely resembling the English hollyhock as both plants belong to the same mallow family. The flowers of the native Indian plant *Gossypium arboreum* are yellow when they first appear, then turn pink, finally red, before withering. As its name suggests Indian cotton grows into a tree-like shrub as high as 4 metres or more. Marco Polo saw this species growing when he visited India in 1290 A.D. and wrote: 'The cotton trees are of great size growing six paces high and attain an age of 20 years.'

When the flower withers the boll bursts exposing the white cotton fibre or 'lint' which has a woolly appearance. This led to cotton at first being confused with wool. Pliny the Elder described India's cotton plants as 'Trees that bear wool'. Much later, an English traveller, Sir John Mandeville returned in 1350 with a most fanciful description of the cotton plant, obviously gathered from hearsay and coloured by his own imagination. He wrote 'there grows there a wonderful tree which bears tiny lambs on the ends of its branches. These branches are so pliable that they bend down to allow the lambs to feed when they are hungrie.' Confusion of cotton with wool lives on in the German word for cotton *Baumwolle*, often seen on clothing labels, literally 'tree wool'.

The length of the cotton which bursts from the boll, the 'staple', varies from 2.5 cm to 6.5 cm. The longer the staple the better the quality of the cloth. The most common Indian native cotton types (*G. arboreum* and *G. herbaceum*) have short to medium staples, but long-staple cotton does also seem to have been cultivated in India from early times. The name for this was *percallas*; we call it percale cotton and find it in high quality sheeting. Today in India, which is one of the largest producers of cotton in the world, all types of cotton including hybrids are grown and high quality long-staple cotton is fairly widely cultivated to meet world demands.

Early Spinning and Weaving

It is thought that spinning and weaving could go back to prehistorical times and that spinning probably preceded weaving with primitive people twisting grass fibres together to make cords strong enough to pull heavy objects or to tie sharpened stones to pieces of wood to make weapons or tools. Weaving is thought to have evolved from primitive intertwining of reeds and fur strips, with a much later development of two basic types of loom, one with weighted warps suitable for weaving flax and wool, and the other a two-barred loom for weaving cotton. The finding of warp weights, judged to be up to 10,000 years old, in silt of the Swiss lakes gave firm evidence of the existence of the first type of loom at an early date. It seems reasonable to suppose that in India both spinning and weaving go back at least 10,000 years for cotton and wool. No doubt the earliest method of spinning used in India was with the fingers. Cotton is especially difficult to spin because of its short lengths, and needs great skill. It is a job that seems to have been done traditionally by women, in India and elsewhere. Inventions throughout the ages have marked progress in efficiency and speed in spinning. The first was the spindle, used from earliest times in India, with its companion, the distaff. Much later came the Indian spinning wheel, the *charkha*, which probably originated in China. The Chinese later improved on this with the addition of a foot treadle. The spinning wheel did not reach Europe until the 14th century when it was used for wool; cotton spinning was not properly mastered there until the mechanical inventions of

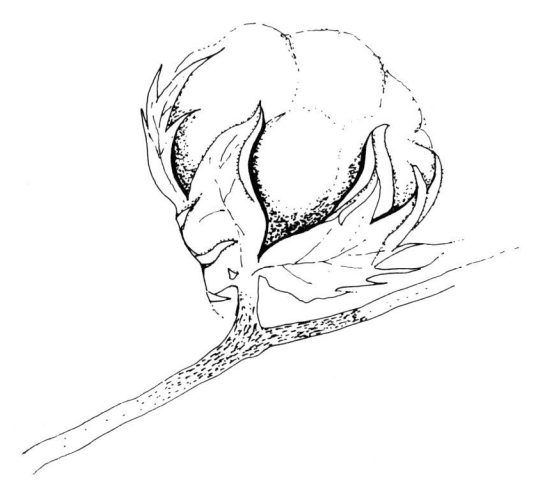

When the flower withers the boll bursts exposing the white cotton 'lint' which is picked and processed. (Artist: Susan Griffiths)

10

A weaver at a traditional upright loom in a present-day handloom factory in south India. It is a scene that would not have changed over the centuries. (Courtesy The Commonwealth Trust Handloom Factory, Calicut, Kerala, south India)

the 18th century. The Saxony wheel was developed in the 15th century based on the Indian spinning wheel but with the addition of Leonardo da Vinci's 'flyer', an invention which twisted the yarn and controlled its continuous winding on to the bobbin. These methods provided the principles of modern mechanical spinning.

Little is known of the earliest cotton loom but it is considered that by the time of the civilisation at Mohenjo-Daro the pit-treadle loom, with heddle harnesses, reed and treadles, or a simple version of it, was in use. Traditional looms are both vertical and horizontal and have varied according to regions. The handlooms used today all over India are still much the same and provided a basis for modern mechanical looms. The spinning and weaving inventions of the Industrial Revolution in England — first the Spinning Jenny invented by James Hargreaves in 1765, the water-frame by Richard Arkwright in 1768 which improved on the Jenny, the (Samuel) Crompton Mule Spindle patented in 1779 which further improved spinning by being able to produce a thread strong enough to weave calico, and the steam engine invented by James Watt in 1769 which powered the machinery and speeded up production far beyond that of hand spinning and weaving — did not improve on the quality of the cotton cloth that India had been producing for so long.

Weavers in India have always been regarded as key craftsmen with their knowledge passed on from father to son for countless generations. Despite invasions and other disasters, weaving and dyeing survived through the family system. Weavers were relatively mobile so that when they fled they could take their knowledge and basic tools of trade with them and easily set up looms under the trees. The Aryans treated weavers with respect. They earned good money and soon began to wield power. This continued until well into the Maurya dynasty, 321–184 B.C., when weavers had their own guilds, the forerunners of unions, to which members made contributions and paid fines, making them well-off and powerful institutions. They could lobby their local government against unfair taxes and could argue strongly with entrepreneurs and merchants against exploitation. It is recorded that a silk-weavers' guild donated money for the building of a temple, acted as bankers, took deposits, and lent money at interest to merchants and others. Inspectors were appointed to ensure that growers, spinners and weavers did not cheat.[4] A system of wages was worked out and bonuses were paid to weavers to encourage increased production. These bonuses were not monetary but took the form of gifts of oil, cakes made of dried fruit which were considered great delicacies, perfume and garlands of flowers. Taxation was based on weight of cloth produced.

Humble Weavers

Weaving was considered to be men's work, but certain women were allowed to work in the industry. These included widows, real ones as well as the 'grass' variety whose husbands were away for long periods on business trips, young women who were cripples and had to support themselves, nuns and ascetics, women who had committed offences and owed fines, retired royal servants, retired temple prostitutes and mothers of prostitutes. Women who had chosen never to marry did not qualify, nor did respectable married women who were expected to stay at home and look after the family.

Top weavers could lead a good life and enjoy a degree of social status. But, generally, weavers, though essential craftsmen, were reduced to being poorly paid members of society, their caste regarded as humble, and this is still the same today although cloth is, as ever, one of India's biggest money earners. Weavers were much more appreciated during the Mughul era (14th to 17th century) when the export of cotton cloth and other desirable commodities made India a rich country and the standard of living was higher, for example, than Elizabethan England. It is ironical, therefore, that within less than 200 years the position was to be reversed, with Indian cotton goods helping to make England one of the richest countries in the world and India reduced to virtual subservience.

This would happen after the Portuguese mariner Vasco da Gama found the sea route to India via the Cape of Good Hope and opened up the mysterious East to Western Europe. In due course India's wonderful painted and printed cottons, the 'Chints', would be on their way to Portugal, then England, Holland and France through the East India Companies, to create a sensation wherever they went and take the various markets by storm.

Notes

1 From *A Note on the Early History of Cotton* by A.N. Gulati and A.J. Turner, 'We may sum up the characteristics of the fabric as being an open cloth, weighing about 2 oz. per square yard, made from 34's warp and weft, and having 60 ends per inch and 20 picks per inch'. — Bulletin No. 17, Technological Series No. 12, Indian Central Cotton Committee (Technical Laboratory), Bombay, 1928.

2 Andrew M. Watson, 'The Rise and Spread of Old World Cotton', *Studies in Textile History*, ed. Veronika Gervers, Royal Ontario Museum, Toronto, 1977, 355-363, Note 1.

. . .The term 'Old World cotton' is used to distinguish *G. arboreum* L. and *G. herbaceum* L., and their ancestors and relatives (which appear to have originated in Asia or Africa), from the possibly more ancient 'New World Cotton', *G. barbadense* L. and *G. hirsutum* L., and their ancestors and relatives (which appear to have originated in some part of the Pacific or in the Americas). Researches of botanists in Russia, America and England during the 1920s showed that these two families are botanically distinct; the Old World species are diploids with thirteen chromosomes (2n = 26), while the New World species are amphidiploids with twenty-six chromosomes (2n = 52). They can be crossed only with great difficulty. See S.C. Harland, *The Genetics of Cotton*, London, Jonathon Cape, 1939, p. 42f; and G.S. Zaitzev, 'Un hybride entre les contonniers asiatiques et américains: *Gossypium herbaceum* L. et *G. hirsutum* L.', *Revue de Botanique Appliquée*, vol. 5 (1925), pp. 628–629.

3 M.D.C. Crawford, *The Heritage of Cotton*, New York, 1948.

There were, in ancient times, two varieties of cotton in Peru; one with a fine white lint of fair length, averaging from 1.25 to 1.5 inches, high in grade and excellent in character, the other a shorter, rougher, more uneven type of a reddish brown color. There is a shadow of evidence that this latter was an older type, since it shows less of the fine points of careful breeding . . . These great horticulturists kept the two types of cotton plants distinct and . . . this distinction exists today . . .

4 The rules were set out in the *Arthashastra*, a government and economic guide written by the brahman Kautalya, adviser to the first Mauryan King Chandragupta.

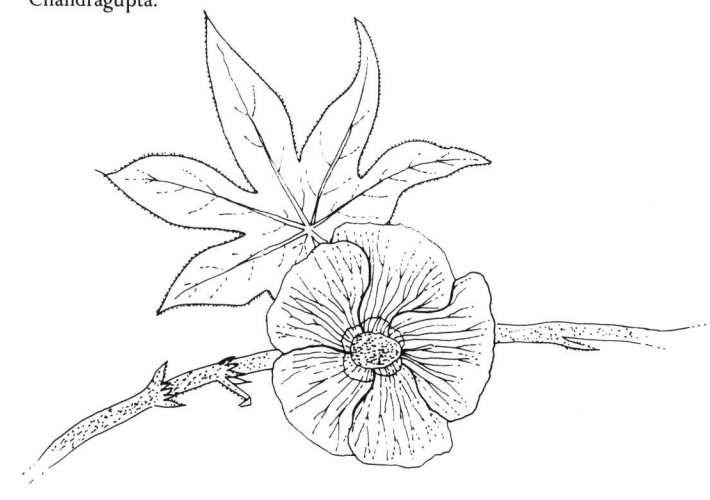

Flower of the indigenous Indian cotton plant Gossypium arboreum *is yellow at first, then turns pink, finally red, before withering. (Artist: Susan Griffiths)*

3. Chintz – So English

Chintz always sounds so English, the essence of traditional English country house decorating. So it comes as a surprise to find that chintz is just as traditional and just as popular in other European countries, particularly in Holland and France where its development has run parallel with that of England. Chintz is also traditional in America where it has been used since early colonial days, imported from England at that time. Also it is often a surprise, even a shock to some, to discover that the word 'chintz', so English-sounding, is actually an Indian word, a variation of the Hindi *chint* or *chitta* meaning 'spotted, variegated, or sprinkled all over'.

Chintz and plain dyed calico first came to England from India as saleable commodities in 1619 imported by the English East India Company and were sold for household use as curtains, covers and bed furnishings. The Portuguese had brought some Indian cottons including chintz to their own country in the 1500s, but they were really not known in Europe until the East India Companies, first the English, then later in the century the Dutch and French, began importing them to their home markets. The Dutch called the patterned cottons *sits*, and still do, and the French called them *toiles peintes* or *indiennes*, and the latter term is still sometimes used in France. It was the English traders in India who used the term 'chint' (singular), plural 'chints'. In early invoices both 's' and 'z' appear. 'Chintz' has now become the universal trade term, not only for high quality flowered furnishing cotton, usually glazed, but for plain glazed cotton. Most unglazed mass-produced printed cottons are usually referred to simply as 'prints'. Glazing was used by the Indian master dyers to bring up the colours and make the cotton gleam like silk, and this practice has become part of the chintz tradition.

It was by no means new for India to be exporting chintz to the West. From several centuries before Christ traders had taken chintz by sea to markets in Arabia, Egypt and Africa, or along the old caravan route through Afghanistan to Persia. The Romans bought chintz. In India whole villages were employed in producing patterned cottons. The Greeks discovered chintz when Alexander's observant admiral Nearchus who had noted the cotton plant reported seeing flowered cotton fabrics called 'chints' that 'rivalled the sunlight and resisted washing', which was a poetic way of describing fast dyes. Earlier than that, around 400 B.C., a Greek physician Ktesias, visiting Persia, wrote of the 'flowered cottons emblazoned with glowing colours which were much coveted by fair Persian women and imported from India'. The Persian word for these flowered cottons was *chitsaz*. Another early reference to chintz is in St Jerome's 4th century A.D. translation of the Bible which likens the lasting value of wisdom to the permanence of the dye colours of India. It is thought also that Biblical Joseph's Coat of Many Colours may have been made of Indian chintz, or a patchwork of Indian dyed cotton pieces. All visitors to India marvelled at the wonderful dyed textiles. Marco Polo recorded seeing the dyed cottons of Gujarat and chints of Masulipatam.

From Gujarat chintz was exported to Indonesia where it was called *tjindai* or *chindai*. The Javanese ordered readymade patterned *tappes* or skirts from India known as *tappechindaes*. In modern Javanese *chindae* or *tjindai* still means 'flowered cloth' but refers to flowered ikat or tie-dyed cloth. It is fascinating to contemplate chintz going West and evolving into the pretty flowered fabrics used today for loose-covers in charming English houses, and going East and becoming traditional cloth for Indonesians. Indian chints went to Cambodia, Thailand, on to China, and later, through the Dutch, to Japan. Another term for chintz used throughout the whole Eastern area was *sarassa* probably from a Gujarat term *saras* meaning 'high quality'. Wherever these patterned cottons went they have influenced textile design and techniques.

The earliest pieces of Indian patterned cottons that now exist were found in Egypt. Some are dated as 14th century and were found in Arab tomb excavations at Fostat, once a busy trading centre near Cairo. Dry conditions enabled the pieces to survive.

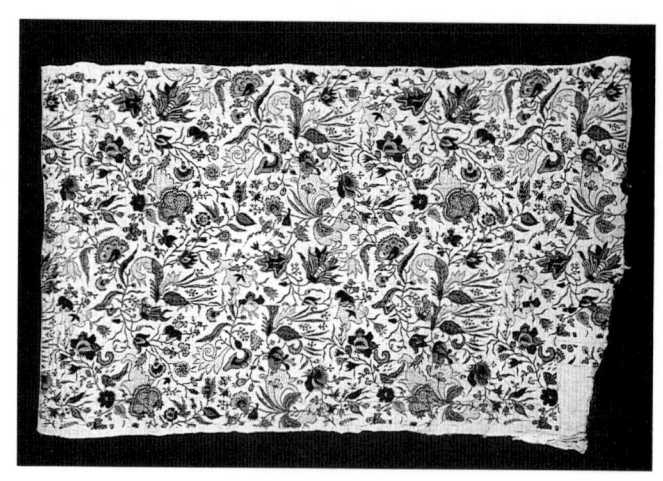

Chintz produced by kalamkari method — drawn resist, painted mordant, dyed; the bright colours of red, purple, blue, yellow, green are in the Dutch taste. This fragment of a skirt, late 18th century, shows the dyer-artist's skill in intertwining flowers and leaves to create a flowing design. (© 1993 Indianapolis Museum of Art, The Eliza M. and Sarah L. Niblack Collection)

Arrival of the Europeans

India was opened up freely to Europe after Vasco da Gama landed on the southwest coast in 1498, at Calicut, the town which gives its name to 'calico'. This part of India, the Malabar Coast, which had been a textile area for centuries was a natural landing point for Arab sailing ships coming from both north and south across the Indian Ocean, catching the monsoon winds. It was the Romans who worked out the use of the monsoon to speed the goods which they imported in such large quantities. The wind would blow the ships one way up towards the Red Sea during six months, then change around for the return journey during the next six months. When the Portuguese arrived the Mughuls were firmly in power in the north, there were powerful Hindu rulers in the south, and India was extremely prosperous with a large export trade in cotton, silks, indigo, and spices. Calicut had a separate ruler, the Zamorin, who was also benefiting greatly from this trade. The Portuguese soon acted to grab the trade by using force to oust the Zamorin. Thus they gained a firm base in India and were the first of the European traders to discover and profit from the country's wonderful products. Among these were the painted cottons which they called *pintadoes* from their own word *pinta* meaning 'a painting', and took back to Portugal. They particularly liked the one-off painted cloths such as palampores (a corruption of the word *palangposh* meaning bedcover) with striking Tree of Life designs featuring exotic flowers and fruits, birds and animals. It was pieces of this type that were first seen in England brought back by early visitors to India, and they were marvelled at for their glowing colours and fantastic designs. They were rare and expensive, acquired by collectors, and embroiderers who sought them to copy.

Thus Indian patterned cottons were not entirely unknown when the English East India Company was formed by a group of London merchants in 1600 with a Royal Charter from Queen Elizabeth I. The Company's first ship sailed east in January 1601.

Their main intention was to trade in spices as the Dutch, who had beaten them out by about five years, were already doing so profitably. They particularly wanted to buy pepper as the Dutch had recently put the price up. The Company also had the idea of selling English-manufactured goods, including woollen cloth, to this large new Eastern market, and this was attempted without much success. The Mughul rulers and their courtiers in India did buy some — they had a liking for scarlet broadcloth — but there was never a big demand. All attempts at selling English products failed, so the Company now turned its full attention to the spice trade.

There was a great deal of money to be made by buying spices from the East Indies (Indonesia) and selling them in the Middle East and Europe, but like all traders before them the English were soon to discover that Indonesian spices could only be obtained by bartering in Indian textiles which had to be of a particular type. Cloth for Indonesia came both from

Gujarat in the northwest of India, where the much sought-after Patolas were made (see 'Design — Ikat') and from the Coromandel Coast. The Indonesians wanted plain and painted calicoes, as well as woven checked and striped cloths—ginghams. They were fussy buyers and required their cloths to be exactly right as to width, weave, colour and design, even the selvedges had to be exact, otherwise they would reject whole shipments. They had been getting what they wanted for centuries and these newcomers had to comply with their requirements. It was a challenge for the English buyers as it had already been for the Dutch.

Factories Established

To buy the textiles in India the Company had to pay in gold bullion, and were criticised at the time for taking wealth out of England for uncertain return. The history of British involvement in India was now beginning. As time went by a firm foothold was gained, often by force, on the Indian subcontinent. Factories were established at Surat on the northwest coast in 1612; at Madras, 1640, and other textile areas of the Coromandel Coast such as Masulipatam; and towards the end of the century at Calcutta (the name is a corruption of Kalikota, the village on the Hoogli River where this great city now stands). From these bases the Company traded in cotton piece goods and yarn, spices and sugar. They bought from established sources through a series of Indian brokers and merchants, but later set up their own weavers working under supervision or on contract and paid advances. A contemporary description shows how primitive were the weavers' working conditions: 'The weavers' houses are under the shade of tamarind and mango trees, under which at sunrise they fix their looms and weave a variety of very fine *baftas* and muslins.' (*Baftas* = calicoes.) Weavers brought finished work to the Company's warehouses, called *kottahs*, to be paid. A maximum price was set with a sliding scale downwards. They were paid according to the agent's assessment of the standard of the finished goods, which could also be totally rejected, or 'turned out', if considered unsuitable. This system was open to abuses and excuses and made it difficult for weavers whose livelihood depended on their efforts, and who had many costs involved in production of the cloth.

The weavers, however, did often turn the rejections to good account by selling the cloth to European and Indian private traders. If the Company had already paid advances to the weavers they were not pleased when rejected cloth, sometimes deliberately made unacceptable, was sold at a good price to outside sources. One method of stopping this practice was by putting the Company stamp on cloth while in the loom, or in the process of being painted or printed, and in due course closing all loopholes. Old chintzes and palempores often show the EIC stamp and date. The English Company found the weavers independent-minded and difficult to deal with. When thoroughly displeased with the Company's authoritative treatment they would take collective action and move themselves from the English area to other weaving settlements such as the French in Pondicherry or the Dutch.[1]

The Dutch, who had been responsible for ousting the Portuguese and stamping out their influence in India, had factories in the same areas as the English so naturally there was fierce rivalry. However, the Dutch did not attempt to make political or cultural contacts with the locals as the English were beginning to do; their main objective was trade, procuring textiles to exchange for Indonesian spices, and to buy Indian spices. Later they did take Indian cottons to their home market and also traded them to Japan, resulting in some Japanese design influences coming into Indian painted cottons.

At first the English Company bought textiles solely for the spice trade, although as early as 1609 a far-sighted agent, William Finch, had recommended to the directors that plain dyed calico for sheets and pintados (chints) for quilts and hangings would sell well in England. But it was to be ten years before these types of cotton piece goods were tried on the home market. At first Indian designs were found to be too exotic for English taste, and the colour combinations were not acceptable. To obtain designs and colours that it was thought would sell the Company officials sent out patterns, or musters, for the Indians to copy. All-over trailing floral designs with light backgrounds had been popular in England in embroideries, and, because there was a vogue at the time for the oriental look, 'chinoiserie' was introduced into the designs. Even the classical Tree of Life design, which dates back to ancient Babylon and is sacred to Hindus as a symbol of fertility and eternal after-life, was affected by European taste and fashion. The Indian tree designs depicting luscious fruit, strange birds and animals on the early palempores were admired but were too 'foreign' to sell well, so the agents requested flowers that appealed to Europeans such as tulip, rose and anemone, and a touch of 'chinoiserie' in stylised chrysanthemums.[2] Also considered to be Chinese in character is the rocky mound from which the Tree of Life is often seen to be growing on 17th and 18th century palampores made for the European market.[3]

The Indian artist-dyers, in interpreting the patterns in their own way, mixing the new ideas with their own stylised naturalistic flowers, produced designs that were different from expected but when they got to the market had immediate appeal. The buying public in England liked the designs and colours but most importantly for the first time had a dyed fabric that could be washed without the colours running or fading in the sun. To have fast dyes was a miracle in those times. In comparison with what they had been used to, linens patterned in one-colour fugitive 'stains', they now had these wonderful cottons dyed in a variety of glowing colours. Another important point was that the prices were low compared with silks and velvets, and thus could be afforded by a larger section of the population. The demand for them grew and grew. The first shipment in 1619 was quite large — 14,000 'pieces' (12–13 yards in length), but a measure of the rapidly growing popularity of the calicoes and chints is that by 1630 around 200,000 pieces were imported. By 1660 enormous quantities were being imported not only by the English but also by the Dutch and French.

The French Arrive

By 1664 the total amount of 'calicoes' imported by the English Company stood at well over a quarter of a million pieces. It was at this time that the French entered the scene. Their company, La Compagnie des Indes Orientales was founded in 1664 and they set up factories at Surat, at Pondicherry on the Coromandel Coast and at Chandernagar, near Calcutta, in Bengal, where the best muslins could be found. The French competed strongly with the English until the mid-18th century when their activities were severely curtailed after a series of military actions won by Robert Clive. French trading activities were confined to Pondicherry which they were to hold as an enclave until the 1950s. However, in spite of their restricted areas the French built up a large and efficient export trade and acquired a knowledge of Indian textiles and a taste for their colours and designs that have made a strong and lasting impression on the European textile industry and have contributed greatly to the fame of modern French fabrics.

In England by the mid-17th century chintz had become the rage. Everyone was buying it. Samuel Pepys wrote in his diary, 5 September 1663, 'Bought my wife a chint, that is a painted East Indian callico for her to line her new study'. Chints were particularly popular for bed-hangings. The various curtains, covers, quilts, carpets, that made up a set for a four-poster bed needed many yards of chintz. The writer Daniel Defoe reported that Queen Mary (of William and Mary) had a bed hung with chintz. He wrote: 'About 1690 at Windsor Castle the late Queen Mary set up a rich Atlas [satin] and Chintz Bed...the Chintz being of Masulipatam on the coast of Coromandel, the finest that was ever seen before that time in England'.

By the 1680s the boom was at its height. More than a million pieces were being imported into England per year and a similar quantity was going to France and Holland. In India, as can be imagined, the activity in the various textiles areas was enormous. Master dyers in India who put emphasis on perfection of their craft resented the exploitation by traders, but money, as usual, was an overriding factor. One reason for the great increase in demand in Europe was that chintz was now being used for clothing. This trend started in Holland where Indian cottons were described as 'being the ware of gentlewomen', but soon were being worn by the whole population. In England the fashion also took on. A Company directive of 1686 stated: 'You may exceed our former orders in chintz of all sorts ... they being become the weare of ladyes of the greatest quality'.

In France *indiennes* were also in great demand. One record states 'As soon as fashionable ladies saw the *toiles peintes* or *indiennes, chits,* etc. they wanted them'. Chintz was made into skirts, blouses, gowns and petticoats. Men wore chintz, too, in vests and dressing gowns. In Molière's play *Le Bourgeois Gentilhomme* written in 1670 one of the characters remarks 'gentlemen wear gowns of *indiennes* in the morning'.

The 'Indian craze' had taken over the whole of Europe. There was consternation in the long-established textile industries — wool in England, silk in France — which were so hard-hit that they successfully lobbied their respective governments to ban the interloper. In England, Daniel Defoe produced a pamphlet deploring the fact that East India chints and painted calicoes, 'India stuffs', were now used for

In every city and village fabric is for sale in shops (bhavans) and street stalls. Here a Rajput cloth merchant in characteristic saffron turban, with his wife, is offering indigo-dyed cotton by weight. (Photographer: Andrew Payne)

Wonderful colours of India seen everywhere in the clothing of the people as in these villagers in Rajasthan. (Photographer: Andrew Payne)

Above: Dutch Wentke of sits (chintz), painted and resist-dyed, Coromandel Coast, 1700–1725. This is a woman's mourning gown; they were made in a sequence of colours from dark blue/black through paler blues to white, finally red. (Courtesy Nederlands Openluchtmuseum, Arnhem, Netherlands)

Right, from top:
Chest protector in bright colours popular with the Dutch. This is an 18th century European print closely influenced by Indian design. (Courtesy Nederlands Openluchtmuseum, Arnhem, Netherlands)

Doll's sleeping bag in chintz, from Coromandel Coast, 18th century. Garments, etc. were patterned to shape in India, ready to sew in Holland. (Courtesy Nederlands Openluchtmuseum, Arnhem, Netherlands)

Modern reproduction of 'Marquis de Seignelay' a textile print originally by C.P. Oberkampf, Jouy, c. 1775, designed by famous textile artist Jean Pillement, with strong Indian design influence. (Document 'Collection Braquenié & Cie' ©)

everything that used to be made of wool or silk. He wrote: 'Above half of the woollen manufacture has been entirely lost, half of the people scattered and ruined, and this by the intercourse of the East Indian trade'. Cotton of all kinds was banned, first in France in 1686, then in England in 1700. It was proclaimed illegal in both countries not only to import Indian calicoes dyed and undyed but also to wear cotton chintz clothing, even the simple versions then being made by the emerging European textile manufacturers. The arrival of large quantities of cotton had marked a crossroads in the development of these new textile industries. In both countries, however, the bans were not entirely effective, although more so in France than in England where a re-export trade of cotton was permitted and means of evading the law generally were soon found. Nevertheless this partial banning did cause a large reduction in imports from India and consequent upheaval among the weavers and dyers with thousands deprived of work and many starving to death.

Dutch Fashions

The Dutch never had a ban and continued to import chintz piece goods throughout the 18th century. They ordered gowns and accessories cut to shape and ready to sew. One traditional dress in demand was the *Wentke*, a mourning gown patterned with stylised flowers on white calico. Mourning continued for up to eight years depending on the relationship of the deceased, and the colours of the flowers went through various stages from near-black indigo, to lighter blues, and so on to various bright colours. Bridal gowns were patterned with very brightly coloured flowers on white. Dutch chintz was always more boldly patterned and more brilliantly coloured than English or French chintz. Samples of chintz in pieces and costumes preserved today in textile collections in Holland, such as at the Rijksmuseum in Amsterdam and other Dutch museums, still show the bold designs and strong colours of 17th and 18th century *sits*. Chintz became an integral part of Dutch regional costume and even palampores were cut up for the purpose. Dutch people still go to auctions to buy Indian palampores when they come up for sale to replenish their regional dress, an increasingly expensive custom as palampores and all old Indian textiles have become sought-after collectors' items.

Cotton chintz may have been banned in England and France but consumers had had a taste of it and their interest was not easily stifled. In France, fashionable women defiantly wore chintz in the face of severe punishments, even death. There were areas outside the law, however, such as the Court of Versailles where the courtesans went on wearing chintz gowns. Another area outside jurisdiction was the papal enclave at Avignon in Provence within which a small textile printing industry had been established, in due course producing 'French Provincial' prints. These were direct copies of Indian block-printed cottons of that time and, continuing to look similar, have gone on being manufactured, unhindered, ever

In Molière's play Le Bourgeois Gentilhomme *Monsieur Jourdain (played here by Roland Bertin of the Comédie Francaise) says: 'Gentlemen wear gowns of* indiennes *in the morning'. This production was brought to Australia in 1988 as France's Bicentenary gift. (Courtesy La Bibliothèque Nationale de France; and photographer: Marée-Breyer, Ivry, France)*

since. The bans were later lifted but things were never the same again for Indian chintz especially after a restrictive heavy duty was imposed on all piece goods into England and France. Chintz continued to be imported but the emphasis shifted to plain undyed calico and processed indigo, both needed for the emerging European textile printers. For a long time the latter had been attempting to produce patterned cottons in the Indian style. They had a certain amount of dye knowledge but their experience was with silk and wool, both of which took dyes fairly easily, and they could not work out how to dye cotton and match the quality of the Indian chints. Even the spinning and weaving of cotton had not yet been properly mastered and would not be until the invention of machines. There was urgent consumer demand for better products. Dye chemists were anxious to discover the secrets of Indian cotton dyeing, secrets that had been closely guarded.

Letters to France

Eventually it was two Frenchmen who found out the techniques of painting calicoes (chintz) and sent reports back to France. These reports, both long and detailed, are now documents famous in the dyeing world. M. de Beaulieu, a naval officer, sent his account from Pondicherry around 1734 to a dye chemist in France. He had witnessed the dyeing process and included samples of chintz cloth at each stage of dyeing with his notes. These are now preserved in the Musée Nationale d'Histoire Naturelle, Paris. However, it is a later and more detailed description of step-by-step methods sent in a letter by the Jesuit Father Coeurdoux in 1742, also from Pondicherry, which has been most often quoted. He wrote down the methods as described to him by dyers who were Catholic converts, and sent them back in a letter to his mission headquarters in Paris. The letter was published in France a year or two later in a collection of letters 'Edifying and Curious' from foreign missionaries.[4] This letter revealed the precious secrets of making dyes fast to cotton and explained the traditional methods of painting the designs on the cloth, known as *kalamkari* (lit. *kalam* = pen, *kari* = work). All the dye substances used, and plant sources of the two basic dyes — madder and indigo — were given (see 'Dyes' for a summary of Father Coeurdoux's letter). It was odd that such potentially commercially valuable dye secrets should have been published, perhaps naive, but as the cotton printing bans were still

imposed in France it may have been thought that the information would be regarded as nothing more than 'curious'. Certainly the ban prevented wide use of the secrets but dye chemists must have taken good note of them because after the bans were lifted in 1759 French cotton printers obviously benefited greatly from the knowledge, gaining a tremendous lead in textile dyeing and in the industry generally, a lead that they have never quite lost.

An outstanding French textile printer who took advantage of the dyeing methods was Christophe-Philippe Oberkampf who founded a cotton printing factory at Jouy-en-Josas, a small town near Versailles, in 1759, after the cotton bans had been lifted. Here he quickly established a market for block-printed chintzes of such high quality that they were greatly in demand by fashionable French women, especially those at the nearby Court of Versailles. Madame de Pompadour, mistress of King Louis XV, wore gowns of Jouy chintz. Oberkampf based many of his designs on the imported *indiennes* with their vividly coloured stylised flowers. He also employed leading artists as designers. His top designer was Jean-Baptiste Huet, one of France's most gifted artists. In this way French design influences such as rococo with its ribbons and stripes became delightfully mixed with Indian naturalism. Oberkampf was known as an inspired genius. Not only was he technically highly skilled but he had great organising ability and business acumen. He was far-sighted and took advantage of every new invention. For instance, he adopted the method of copper-plate printing of cotton which had been invented earlier in Britain. His 'plate' prints — one-colour engraved scenic prints — became so famous that 'toile de Jouy' became, and remains, the generic term for textile prints of that kind wherever manufactured.

From the middle of the 18th century other textile factories in Europe were producing chintz of varying degrees of quality. There was excellent work being done at factories in Mulhouse, a town on the Swiss border, which was well-established as a textile printing centre. It had been unaffected by the French bans being at the time under Swiss control; it was here that Oberkampf first learnt his trade. The English industry was also developing. A talented textile designer was William Kilburn who had his own calico printing factory in Surrey in the 1780s. He produced exquisite floral designs of great originality printed to the highest standard. His cloths were expensive and, unfortunately, in spite of legal protection of his designs, cheap imitations were made which undercut his on the market. He went bankrupt, but not without leaving a lasting influence on English chintz design. The Dutch had also founded their printed cotton industry which today is one of the world's largest.

Discovery of 'Mauve'

Indian dye methods were slow and laborious but they provided European dye chemists with the clues which led to the invention of synthetic dyes. There were many advances in dye chemistry in the late 18th century. A Swedish chemist, C.W. Scheele, discovered chlorine as a bleaching agent in place of lemon juice and sun which the Indians had used. Then, in the mid-19th century William Perkin, an English chemist, in trying to synthesise quinine, accidentally discovered the dye mauveine. The French called it *mauve* and it became fashionable all over Europe accounting for all the lavender gowns of 19th century literature. In 1869 the essential ingredient of madder, alizarin, which provided Indian dyers with reds, was synthesised by two German chemists. The greatest European dye triumph, however, was the synthesising of the Indians' important blue dye indigo by another German chemist, Adolf von Baeyer, at the end of the 19th century. He was awarded the Nobel Prize for this discovery which was a measure of the importance of these dyes. Until then natural indigo had been grown in India, and in the West Indies since the early 17th century.

Many of the early synthetic dyes were of poor quality, not fast, and crude in colour. This, combined with the mechanisation of printing brought about by the industrial revolution, tended to produce cottons that were badly printed, and also of uninspiring design, particularly in the mass market. One man who sought to improve things in England was William Morris who led the Arts and Crafts Movement which promoted a return to high quality craftsmanship and design. Morris used the old method of wooden block-printing and took his inspiration from medieval designs. His products were admired not only in England but all over Europe, but because the manufacturing process was slow they were too expensive to compete on the market. Later, when techniques improved, Morris did use machine-printing and his designs in fabrics and wallpapers have endured to this day. There were several British firms established around the same time which were producing high quality printed textiles and some are still in existence (see Appendix 1: 'Fabrics for Restoration').

It can be seen that the 'spying out' of the Indian dye methods and the copying of their designs led to enormous economic advantage and textile supremacy in the West. It also dealt a crippling blow to the livelihood of Indian weavers and dyers, and to their morale. They not only lost the precious secrets which they had guarded so carefully but also to a large extent the recognition of their genius, recognition that has really not yet been regained.

Notes

1 From the *Indian Economic & Social History Review*, July–September 1980, Vol. XVII, No. 3. 'Weavers, Merchants and Company: The Handloom Industry in South-eastern India 1750–1790.' S. Arasaratnam, pp. 257–281.

2 The chrysanthemum is actually of Indian origin but went to China where it was held in high esteem, celebrated by Chinese poets. It also went to Japan where it became the national flower and has been developed in variety, excellence of quality, form, colour and size. — 'Flowers in Indian Textile Design', Vijay Krishna, *Journal of Indian Textile History*, VII, Ahmedabad, Calico Museum of Textiles, 1967, 1–19.

3 From 'The Flowering Tree in Indian Chintz', K.B. Brett, *Journal of Indian Textile History*, III, Ahmedabad, Calico Museum of Textiles, 1957, 45–57.

4 Published in the 26th collection of *Lettres édifiantes et curieuses écrites des missions étrangères par quelques missionnaires de la Compagnie de Jésus*, Paris, 1742, and introduced as a 'book suitable to amuse and edify in piety'. — From *French Documents on Indian Cotton Painting*, Chapter II, 'New light on old material', P.R. Schwartz, Studies in Indo-European Textile History, Ahmedabad, Calico Museum of Textiles, 1966, 94. A second letter with further explanation was written in 1747.

4. All the Colours of the Rainbow

If cotton is India's greatest textile gift to the world colour cannot be far behind. Not only did Indians invent fast dyes but they produced colours which reflected nature in all its vivid glory. Everywhere you go in India there is colour. It is apparent in everyday life as well as in festivals and processions, in the temples, in flowers which play so much part in ritual, but particularly in the clothing of the people. There is every conceivable colour in the saris of the women; men, too, wear colours, in turbans, sashes, and shawls in contrast to their white shirts and trousers or dhotis; and saffron and magenta-clad monks thread in among the crowds in the busy streets.

The colours of India are really wonderful. They are incredible hot pinks, brilliant oranges, vibrant reds, purples and mauves, every shade of green from palest lime to deepest emerald, peacock blue, turquoise, yellows, corals, all seeming so modern, as if recently created, yet in truth centuries old. Here is a pretty description by a 7th century A.D. poet of preparations for the wedding of a princess in northern India:

'The palace was arrayed in textures flashing in every circle like thousands of rainbows, textures of linen, cotton, bark silk, muslin and shot silk' . . . all dyed and spread out to dry before being made into costumes for the wedding. The passage concludes: . . . 'all these gave to the court an aspect brilliant, attractive, exciting and auspicious'.[1] Then, as now, a variety of beautiful colours and textiles added quality and richness to life in India.

Indians have an instinctive colour sense that enables them to mix colours with subtlety and balance, but also with a touch of exciting discordancy. For centuries they have been putting together blue and green, pink and orange, purple and green, red and coral, all daring combinations that once would not have been tolerated in the West but are now considered high fashion. Schiaparelli may have thought she invented shocking pink but a visit to India would dispel that idea. You see it everywhere, especially in the desert states of Rajasthan and Gujarat.

Subtle Shades

Though they love strong colours and contrasts Indians are also sensitive to subtle gradations of colour. Kamaladevi Chattopadhyay points out that white, for instance, has five tones with poetic classifications — ivory, jasmine, August moon, August clouds after rain, and conch shell — all delicate and gentle. Fading never bothered Indians. As with all natural things fading was accepted as inevitable; just as flowers faded so textiles faded as they aged, giving a gentle, mellow look. A woman might have her sari deliberately done in fugitive dyes so that when it lost its colour through washing and drying it could be re-dyed in a completely new colour. Throughout the country in cities and villages it is still easy to go to a local dyer and have a sari re-dyed, or a new one dyed in colours and designs to suit. Traditionally, colours were never based on fashion but were influenced by religion, regional climate, geography and natural surroundings. Colours reflected the sky in its various shades from dawn to dusk, even the night sky, water of rivers, lakes and sea, flowers and vegetation, and the various shades of stones and earth. Colours were also dictated by such practical considerations as available plants for dyes and the local water being suitable for dyeing. The result, as eminent Indian textile historian Pupul Jayakar notes in her book *Handloom Textiles* (Bombay, 1973), is a colour wheel around the country.

The wheel begins in Rajasthan, the desert state in northwest India, where the colours are vividly bright against the barren, stony background. It then turns down west to Gujarat, also semi-desert, known as the textile state and renowned for wonderful colours in its famous tie-dyed and block-printed fabrics, and continues towards Bombay. The colours begin to moderate in the hilly country, then grow paler until they reach the lush tropical regions of the Malabar Coast where men wear gleaming all-white and women wear white or palest pastels leaving nature itself to provide strong contrasts. The paleness continues to the tip of India, Cape Comorin, where the wheel turns up the east coast back into barren lands where bright colours once again prevail, up the Coromandel Coast, traditionally famous for *kalamkari*, continuing into the states of Andhra Pradesh and Orissa, both noted for colours as well as for textures and patterns. The colours now grow paler as the wheel travels north via Calcutta and West Bengal into Bangladesh, formerly East Bengal, another 'textile state', ending as white and pale in the fertile green tea-growing regions of Assam. In the centre of India, on the open cooler plains of the Deccan, colours are earthy such as maroon, chocolate and dark red contrasted with yellow and white.

Colour symbolism plays a significant part in the Hindu religion which means it affects the lives of the majority of Indian people. Colours of good omen predominate and red and yellow are the most important. Red represents energy, joy and happiness and is worn at weddings. Red is also worn when the first monsoonal rain falls as a sign of rejoicing after months of relentless heat and for the promise of renewed fertility of the land. Yellow, saffron yellow, is a religious colour, the colour of renunciation and sacrifice. At weddings while the bride and her entourage may wear red the bridegroom will wear a handsome white suit with saffron turban, sash or shawl. Red and gold are also colours of splendour and ceremonial, as they are in the West, associated with royalty and richly decorated palaces, with great durbars and public parades.

Festivals of Colour

At religious festivals there is a predominance of yellow, with yellow flowers such as marigolds or yellow hibiscus, a highly esteemed flower, piled around statues of goddesses in processions or in the temples. There is one festival that has many colours. This is Holi, a fertility festival when rainbow-coloured powders — red, purple, pink — are thrown at people followed by a squirt of water. Everyone wears light-coloured washable clothes. Children love Holi because there is complete freedom to throw the colours at anyone. There is dancing and music, general excitement, and everyone is happy. Another popular festival is Divali, the Festival of Lights which fights off evil spirits. Thousands of twinkling lights are strung up everywhere, outlining buildings, and flickering inside houses. Fireworks are part of the festival. The people dress in their best clothes, the women and little girls wear shimmering shiny fabrics, and all flock to the street temples which are ablaze with light and colour.

Colour is traditionally associated with the four great classifications of caste. White, pure and sacred, belongs to the Brahmans at the top of the caste scale; white also represents their contemplative characteristics. Red and saffron are for Kshatriyas, the warriors (Rajputs) and aristocrats; the red of the Rajput signifies the essence of vitality and vigour. Green which represents regeneration and production is for the Vaishyas, to which belong manufacturers and merchants of the middle class. Blue or black, both colours of ill-omen, are relegated to the Sudras, workers and servants. At the bottom of the scale are the Untouchables, or Harajans, for whom no colour seems to have been designated. These are the main castes; there are thousands of sub-castes or sects. Weavers belong to the Khatri sect, and dyers are an off-shoot with various sub-sections. Castes refer to traditional occupations. In the modern world people in a variety of occupations such as academics, professionals, government servants, journalists and so on would come from many castes. Even in ancient times before castes became clearly defined colours were recommended for different sections of the community. 'Householders', namely reputable gentlemen, were advised to wear white because it was cool and practical. Most men in India today wear white for the same reasons. Widows were required to wear white as a sign of mourning but a Hindu woman whose husband was alive could not wear white, or black, another colour of death. Black was never liked, even prohibited. These mourning rules are still followed especially in the upper castes. One meets, usually in enclosed family circles, dignified old ladies wearing exquisitely embroidered all-white muslin or lawn saris. In some regions black is prescribed for mourning. Apart from the black/white rules women were always allowed freedom to wear warm auspicious colours. There does seem to have been ambivalence about blue. On one hand it was regarded as a colour of ill-omen, on the other as a colour of the sky, of infinity and associated with the god Krishna. It was also regarded as a colour that was becoming to women, which is probably why it later fell into disfavour with Brahmans. Because blue made women look attractive it became associated with human love, a condition which prevented a man from concentrating on a religious life such as becoming a priest. By pronouncing blue to be a colour of sorrow and ill-omen Brahmans aimed to discourage the wearing of it and keep men under their control.

The caste colours given are generalised. Colours vary according to clans and regions and are influenced by local caste history and tradition. Just how complex these colour traditions can be is illustrated by the rules of one of the Rajput clans, the warriors of the northwest desert states. The intricacies were explained by a Rajput prince, Maharajkumar Jayasinhji Jhala, who is an anthropologist, and a son of the Maharaja of Dhrangadhara, a state in far western Gujarat, on the Kathiawar Peninsula. Although the prince's father has lost his power and state income the royal palace is still a family home with life functioning to meet practical modern needs. His mother, the Maharani, and an aunt used to live there and a younger brother, MK Sidhhrajsinhji, still lives there with his wife and two children. This prince gave up a career in France to return a few years ago to set up a craft workshop based at the palace. Here he has revived the silver, stone, wood and glass craft traditions of the area and employs many people. The beautiful products are sold in Bombay and other Indian centres, and overseas. The palace zenana is still inhabited by a few old servants who have lived in these traditional women's quarters all their lives.

The people of the kingdom continue to honour their royal family and when there are special gatherings the prince may wear a splendid turban in the state colours of watermelon pink, navy blue and gold. The pink represents the dawn and dusk; the blue, the deep night when the founder of the kingdom with his divine consort conquered and created the Jhala kingdom in the 1090s; and the gold represents the divine pair. Only the royal males wear these colours. Different occasions call for turbans of different colours but most characteristic for all Rajputs regardless of status is saffron yellow representing renunciation and sacrifice, values to which the Rajput is born. Rajput death-ritual colours are especially significant. If a Rajput woman dies leaving behind her husband and grown sons this is considered an auspicious parting. She has escaped the worst possible condition a woman can experience in life, that of being a widow; also she has done her duty by her husband, had the honour of bearing the seed of the Rajput clan and borne his sons to carry on the lineage. Her auspicious death has been a reward for her good deeds and she has earned the right to be dressed for her funeral in her wedding clothes of red and green.

Royal Colours

If a royal Rajput woman were in mourning for a member of the family she would wear during the mourning period graded shades of white and black combined with another colour depending on the relationship, time of death and age of the person concerned. For example, if her eldest adult son's wife died, as mother-in-law she would wear black with a dark red border, provided her husband were alive. Later she would wear white with colour in the border and after the mourning period would return to warm auspicious colours. If she were a widow she would wear white for as long as she lived. This is the rule for all Rajput widows, but if she were under 35 and widowed she would go through the black/white mourning phase, then on an everyday basis wear neutral colours in order not to draw attention to herself. A Rajput man would never wear black unless he were exiled from his country. Accuracy in colour symbolism is extremely important, affecting not only religious sensibility but protocol, especially in aristocratic circles. In the zenana, for instance, there was a strict code of colours and types of fabrics depending on status and no one would have dared step out of line.

Other Indian religions have their own colour symbolism. The Jain sect, a puritanical off-shoot of the Hindu religion, relates colour to personality traits. Red, yellow and white are for virtuous people; blue for the great middle range, not so good, not so bad; grey for thieves; black for anyone really cruel or evil. Jains, thrifty hardworking people, are often found among the rich merchant class usually involved with textiles. There are many beautiful Jain women noted for their exquisite taste in tie-dyed and printed silk saris, and choice of jewellery.

Buddhists use colour to define behaviour. White or yellow are for healing, green or red for bewitchment, blue for destruction. Saffron or magenta are their sacred colours. For Muslims green is the sacred colour. A male Muslim who has made the pilgrimage to Mecca is entitled to wear a green turban. Muslims regard yellow as a colour of happiness and good fortune; it is their wedding colour.

Astrology also dictates choice of colour, certain colours being more auspicious than others depending on the stars for a particular person on a particular day or occasion.

With so much formality, so many factors limiting free choice, it is amazing that there should be such artistry and satisfying balance in colour selection, and such an instinctive flair for choosing and mixing the right ones.

Note
1 From 'References to Textiles in Bana's Harshacharita'. V.S. Agrawala, *Journal of Indian Textile History*, Vol. IV, 1959, Calico Museum, Ahmedabad. The princess was a sister of King Harsha who reigned A.D. 606–648 after the main Gupta period.

Major textile areas 1600-1750. Gittinger, Mattiebelle, Master Dyers to the World, *The Textile Museum, Washington, D.C., 1982.*

5. Dyes and Methods – The Ancient Secrets

Most amazing is that Indian dyers had for so long been able to make colours fast to cloth — silk, wool but above all cotton which, with its waxy surface, is resistant to dye. Scholars think that the dye chemistry they used was discovered by ancient alchemists as a by-product of their search for the elixir of life.[1] The art of dyeing was developed from this knowledge with the skills and knowledge passed on from one generation to the next throughout the centuries. It is not known how far back dyeing goes but it certainly dates to the early Indus Valley civilisation (more than 4000 years) as proved by the dye in the cotton fragments found at Mohenjo-Daro and by the discovery of dye vats among the ruins there; cloth dyed in many colours is also mentioned in the earliest literature such as the *Mahabharata* which chronicles mythical events going back to circa 3000 B.C. Specialised techniques to create coloured designs evolved over time, such as hand painting (*kalamkari*), block printing, tie-dye resist techniques of ikat, bandhanni and others, and patterning in the loom (brocade, etc.).

Washerman's Work

Dyeing was mainly a male occupation with women employed on the periphery. Women have traditionally done the tying for tie-dyed resist techniques; and when there were huge trade demands for chintz, hand-painted or block-printed, whole villages including children helped with the finishing work. Village dyeing was done by the dhobi/rangrez (washerman/dyer). He washed newly woven cloth thoroughly and bleached it for dyeing. He also did re-dyeing. The *Arthashastra* (c. 320 A.D.) set down rules for the length of time the dhobi/dyer should take to carry out various commissions, as follows:

Cloths to be washed only, to be returned within 1–4 nights; to be lightly coloured, in 5 nights; to be dyed blue with indigo, in 6 nights; to be dyed red, with lac or saffron, or requiring much skill, in 7 nights.

At this early period commonly used dye plants such as madder (red), safflower (saffron) and indigo (blue) had begun to be cultivated in an organised way.

Dyers were less mobile than weavers who could set up new looms by using tree trunks and branches. Dyers were restricted by their equipment which, though simple, would have been difficult to replace easily — vat, vessels for boiling, mortar and pestle, mixing bowls and, if these were ceramic, thus porous, separate ones would be needed for each dye colour. There would also have been attachment to and knowledge of regional plants and other dyestuffs, and the need to be near the right type of running water. Another difficulty was that specialist techniques like painting or printing of cotton involved a team with each craftsman completing a different stage. Unless an experienced group could be kept together moving would be highly disruptive. On the other hand the family apprentice system meant that there were always trained craftsmen ready to step in, and this is still done, with blockprinters and others being able to pick up and leave work for another to take over at any time.

The dyers who gave away the dyeing secrets to Father

Coeurdoux were from the Coromandel Coast, a region long renowned for the excellence of its chintz, for the quality of its cotton and its lustrous dyes. Marco Polo mentioned the 'beautiful chintes of Masulipatam' which was the textile centre of the area. Pondicherry, the French territory where the dyers were from, is about 136 kilometres south. The French priest could not have received his information from a better source.

The Dye Secrets

The following is a summary of Father Coeurdoux's letter of 1742 from a translation by P.R. Schwartz, an eminent French authority on the history of dye chemistry.[2]

First, they chose a good piece of woven calico in its natural or 'grey' state. The best was known as percallas *(percale) noted for fineness and regularity of weave, and woven from long-staple cotton. Then came the steps and recipes which were to provide the basis for our modern dye techniques. The initial preparation of the cloth was most important. The waxy surface of cotton which is resistant to dye was first broken down. To do this the cloth was thoroughly washed, then semi-bleached by being soaked with lemon juice, then dried in the sun, with this step repeated several times.*

The cloth was now soaked in an aqueous solution of fat and astringent (buffalo's milk mixed with the powdered root of myrabolans plant, also known as cadou[3]), wrung out, and the process repeated many times, followed by more sun-drying, washing, and drying in the shade. Next came 'beetling' — a thorough beating, wood on wood (tamarind wood was used in this area) to give the smooth surface needed for painting.

With the cloth now properly prepared came the 'pouncing' of the design. This was done by the artist himself. Execution of the design was considered the most important stage and the artist, who belonged to a different caste from the dyers, was given special respect. He first drew the design on light paper; then the outline was perforated and powdered charcoal dusted through the holes. The charcoal-traced outline was then drawn over with a pen (kalam) made of a small piece of bamboo sharpened and split at the end and dipped in mordant — first acetate of iron, then a solution of alum tinted with sapan wood (madder); the two together gave a good black.[4]

The French priest's account explained that with the outline of the design now drawn the cloth was dipped in a vat filled with red dye derived from chay (a plant endemic to southeast India), the effect being to blacken still further the outlines and develop the red lines. The cloth was then soaked overnight in water to which diluted sheep's or goat's droppings had been added. This solution acted as a bleach because, as the dyers explained, its effect was to rid the cloth of unwanted red dye which, if left, would cause the blue dye when applied to become black. It also further whitened cloth which was only half-bleached at the beginning.

There was further treatment and bleaching of the cloth to prepare it for the indigo. Now the cloth was covered with beeswax except for the parts to appear blue or green. The wax was applied with a brush fitted with metal points, the fluid wax being released from a ball of hair and twisted hemp wound round the stem. The cloth was now dipped in a vat of prepared indigo and left for about an hour and a half; when removed it was found to be dyed in the right places. The wax was removed in boiling water and the cloth again thoroughly washed, beaten and bleached by the use of several processes including a soaking in water containing a small quantity of cow or female buffalo dung, then hung in the sun to dry for days before it was considered ready to receive the other fast dyes.

The next step was waxing the lines required to appear white within the red areas, followed by painting on of the alum mordant consisting of a solution of alum, tinted with sappan wood (gives red). The composition of alum mordant varied according to the tones required. A weak solution gave pink, a much stronger one gave deep red, the addition of iron gave violet. The cloth was again dipped in a vat filled with red dye (chay).

Yellows were painted on using turmeric or saffron, both fugitive; green was obtained by painting a rather complicated concoction of juices from plants and fruits on blue.

As a finish the chints were given a gloss to bring up the colours. The Indian method was starching with rice, then beetling and chanking (burnishing with a shell) until a satin-like sheen was obtained. Traces of starch have been found in old chintzes in museums. The glazing on chintz had a practical appeal in 17th century Europe because it repelled dust. Modern glazing is done by rotary ironing or 'calendering'.

In due course European dye-chemists were able to analyse the dye substances which had been set out in Father Coeurdoux's letter. Apart from myrabolans the main ingredients were madder (*Rubia tinctorum*) which contains the red colouring matter alizarin and was used in conjunction with mordants; and indigo (*Indigofera tinctoria*), the main source of blue dye.

Madder

Indian plants containing alizarin are known as *al*, or *aal*, and also as *saranguy (Morinda citrifolia)*, as well as other shrubs of the genus *Morinda*, all used in west India; *al* is called *chiranjee* in south India.

Another red-yielding plant is *chay (Oldenlandia umbellata)* growing on the Coromandel Coast. This produced the deep rich red for which the area was famous. It grew near the Kistna delta where there were rotting shells in the soil. Indian dyers knew it produced a good red but it took European chemists to work out that this was due to the lime in the soil, so henceforth calcium was added to madder to deepen the colour. Another plant which gives red is sappan wood *(Caesalpinia sappan)* also called brazilwood. It is pounded and used with mordants to give different tones of red and brown, but is not fast to cotton.

The madder plant had been known in Europe but dyers could not get it fast to cotton until they discovered the Indian dyers' use of mordants. The Indian dyers could produce many colours using the mordants alum and iron, in conjunction with madder and other red-yielding plants. Alum combined with tannin used with madder could produce a scale of reds through to pinks. Iron, combined with tannin produced black. Violet resulted from a combination of alum and iron with madder. Pomegranate rinds were also used as mordant. By adding flower petals, leaves and other parts of plants as colouring matter the dyers could achieve many different shades. The recipes and methods given by the Coromandel dyers came from one area and from one group of dyers. They varied from region to region depending on available substances, but there do seem to have been similarities around the country. A medical treatise setting out recipes used by dyers in Hindustan during the Mughal period (1556-1803) transcribed in 1766, lists 36 shades involving more than 90 processes, along the same lines as those used by the Coromandel dyers. Mordants used in Hindustan at this time were alum, iron and myrabolans, and, in addition, lemon rind, green dried mangoes, cotton flower and nuts of *bhalawan (Semecarpus anacardium)*.

There has been a revival of natural dyeing in India in recent years sponsored by the All-India Handicrafts Board in an effort to keep old crafts alive, and a similar project was launched in 1982 in Bangladesh (East Bengal), long famed as a textile region, by the Bangladesh Small and Cottage Industries Corporation. In both cases lists of traditional dye-yielding plants and substances, and guidance for their use have been published.[5] Here are a few of the plants from all areas, which give an idea of what dyers throughout the centuries have used:

Annatto *(Bixa orellana)* — seed yields orange.
Basil *(Ocimum sanctum)* — leaves give lime green.
Cardamom *(Elettaria cardamomum)* — leaves give pink.
Catechu *(Acacia catechu)* — wood extract yields brown/maroon, used for dyeing canvas sails and fishing nets.
Coral jasmine *(Nyctanthes arbor-ristis)* — petals give yellow.
Frangipani *(Plumeria rubra)* — leaves give light olive green.
Henna *(Lawsonia inermis)* — this is a very old dye plant, the leaves of which give the familiar henna/gold.
Jackfruit *(Artocarpus integrifolia)* — sawdust yields yellow ochre/light yellow. Jackfruit grows in the Ghats, also a tea-growing area, of southern India.
Marigold *(Tagetes erecta)* — petals give mustard.
Pomegranate *(Punica granatum)* — rind used to yield khaki, and with iron mordant gives dark grey.
Safflower *(Carthamus tinctorius)* — this is one of the oldest dye plants; the petals produce the yellow/orange saffron seen so extensively all over India in turbans, robes and sashes. Red can be separated from the yellow by treatment with an alkali.
Turmeric *(Curcuma domestica)* — another old and widely used plant; yellow dye comes from the roots.

Mordants used today are alum, tin, chrome (potassium dichromate), copper (copper sulphate), and iron (ferrous sulphate).

Madder with alum gives dull Indian red; with tin, gives orange; with chrome, gives garnet; with copper, gives light bronze red; and with iron, gives reddish violet.

Indigo

This is a legendary dye and was almost as important as cotton in the Indian export trade from ancient times. It is a native plant of India, the name deriving from a Greek word meaning 'the Indian substance'.

In India it is called *nila* (Sanskrit) and the indigo-dyer belongs to the *Nilaro* caste. As a dye-stuff it is particularly suited to cotton, but loses its intensity on linen, and is difficult to use on silk. Indian dyers discovered thousands of years ago how to process it by fermentation and oxidation to produce a vat dye-stuff that could be used without mordants. It has also been used as a medicine; the Romans imported indigo both as a medicine and a dye.

Before indigo went to Europe blue dye was obtained from woad which contains the same dye essence, a glucoside called *indican*, and is processed in a similar way. In due course, in spite of much resistance from the big established woad industries of southern France and Germany, woad was supplanted by indigo which was found to be much more effective as a dye.

The processing of indigo was described by Marco Polo on his visit to Gujarat and it has not changed. 'India has a species of herb which is plucked out by the roots and put into tubs of water where it is left to rot. Then the sap is pressed out which when exposed to the sun evaporates leaving a kind of paste behind it which is cut up into small pieces'. The small blocks thus produced became the trading commodity and were used by dyeworks in England, France and elsewhere.[6]

To activate the dye, fermentation is brought about by diluting the paste with water, then mixing it with an alkaline substance and sugar (an Indian recipe was lime and molasses) for several hours. Fermentation of indigo is one of the most difficult operations of the traditional dyer's craft. To get the dye bath 'just right' a dyer has to rely on luck, intuition and experience. When the bath is ready the cloth is dipped and wrung out several times according to the depth of colour required. When exposed to the air, or oxidised, the colour holds fast, becoming indelible. The Indian dyer kept dipping and wringing the cloth so that his own hands became permanently stained with blue. His status was low, and there was no way of hiding it. In Manchester in the 19th century indigo dyers were known as 'blue-handed boys'.

Apart from staining the hands indigo has a certain toxicity so there was also a danger of inhaling the blue powder. To avoid this the dyers kept their whole faces except their eyes covered.

Dye Superstitions

Because of the difficulty in fermenting indigo many superstitions surrounded the process. Dyers talked about the 'evil eye' being on the vat. Rudyard Kipling's father, J.L. Kipling, curator of the Central Museum, Lahore, during the Raj years, wrote in an article published in the *Journal of Indian Art and Industry*, Vol. I, 1886, 'At Lahore I visited a blue dyer's workshop and was only permitted to inspect the vats on taking off my shoes, the superstition being there prevalent that without that act of reverence the evil eye which might spoil the fermentation would not be averted.'

Indigo provided the Indian dyer with a variety of blue shades from the lightest, known as *baiza* meaning 'crow's eggshell', through sky blue, to the deep blue seen today in denim jeans. The dyers could produce turquoise and greens by the addition of turmeric, or saffron. Used with madder, indigo could produce a scale of mauves through to deep purple. Natural indigo dyeing still takes place on a commercial scale for the domestic market in India, in one area of Andhra Pradesh. The colour produced is black and the dyed yarn is used for the weft of shot-colour saris. The dyers still rely on experience and good luck to control the stages of fermentation and de-oxidation of the dye. They also offer up prayers to the Goddess of the Vat, Ghata Mali, for the success of the operation.

Block-printing

The Indian technique for painting cotton as described is known as *kalamkari*, literally 'pen-work'. Because of the time involved in laboriously painting the cloth (some work took months to complete, even years) and in order to meet the huge demand for chintz from western traders, the Indian dyers also used the labour-saving method of block-printing. It is not known for certain when or where block printing originated. Some sources attribute its invention to China where it was used from the 1st century A.D. for printing silk. Others place it at a much earlier period in India saying that it is mentioned as *chitranta* in the *Ramayana*, the Sanskrit epic. Block-printing was certainly used in Gujarat in the 13th century to meet the big demands of the Arab trade to the West. Many villages around Ahmedabad are still centres of block printing; Pethapur is the best-known one, devoted entirely to block-printing. For block-printing the Indians used the same mordanted dye methods with the addition of a gum to fix the dyes to the cloth. Because of the demand and work being done too quickly it was often slipshod, and shipments were sometimes rejected. However, there are many examples preserved in museums of beautiful Indian block-printed chintzes of the 17th and 18th centuries. The dyers have managed to achieve the same movement and flow of floral designs as are shown in the *kalamkari* cloths.

Block-printing was used by European textile printers until well into the 19th century, and even the 20th by some high quality printers. It was gradually superseded by the invention of speedier methods such as roller printing, and, ultimately modern rotary and flat screen printing.

Revival of Block-printing

Since the early 1960s there has been a tremendous revival of hand block-printing in India with a growing international demand for printed cotton clothing and household linen, such as tablecloths, quilt and cushion covers, and yardages. Block-printing is carried out in various places but particularly in Jaipur, Delhi and in the traditional block-printing villages around Ahmedabad. Workshops vary in size but generally are organised along the lines of the old *kharkhanas* (workshops) with the same hierarchy of craftsmen and a traditional system of work. On the whole it is a male occupation but in Jaipur and Delhi there are some very successful women designers running their own workshops employing male traditional craftsmen with women to do the sewing. In some workshops there are women dyers. More and more modern young women, graduates of textile colleges, are coming into this field.

Mixing the colours is complicated. Best synthetic dyes are used, usually vat dyestuffs as being the fastest to cotton, and the master dyer must mix them to a strict formula as the colours placed on the fabric are no guide to the desired finished colours. These do not appear until the dye has been oxidised at the end of the process. A gum is mixed in with the dye to make it adhere to the fabric.

The designer sketches the motif and the master blockmaker adapts and transfers it to the block. Sometimes the master blockmaker is also designer. Such a craftsman is very talented with an instinctive feel for design and is usually descended through many generations with training in the craft being passed from father to son. A blockmaker is called a *chhipa* (lit. carpenter). He makes the basic block *(chappa)* from hardwood such as teak or *sheesham*, a type of rosewood. To get the right finish he will spend a long time planing and sanding and is not satisfied until the block is perfectly smooth and flat. Most blocks are 8 inches (20 cm) square, with border blocks 3 x 6 inches (8 x 15 cm). Larger blocks are also made, up to 14 inchs (36 cm) square, but the colour tray *(parat)* used by the printer is usually 16 inches (41 cm) square, and this limits the size of the block.

Engraving of the block is done with special tools, the *kalam* and *thasso*, which are chisels in a variety of gauges, from very thin to fairly thick. The design is first drawn on a piece of waxed paper and finished in transfer ink. The paper is placed with drawn design face down on the surface of the block which has been whitened so that the design will show up well. The blockmaster transfers the design to the block by using a fine chisel and lightly tapping holes through the outline. With the design transferred he removes the paper and begins to carve the block following the outline. He may do all the work himself or give some of the preparatory work to his assistant. Intricate carving is a highly skilled job and it takes two to three days to make a set of blocks for one design which includes the outline block and the various colour blocks for overprinting. A master blockmaker will usually make his own tools, or inherit them.

Old blocks retrieved from places like Masulipatam and Ahmedabad are often used for inspiration and reference. They can rarely be used for printing because of broken edges, but it is sometimes possible to repair them sufficiently for limited production. This type of work has appeal to connoisseurs but is not practical for commercial production.

Block-printing Steps

Here is the process for block-printing a tablecloth as followed by a block-printing workshop near Delhi which produces export goods.

1. First, a piece of plain cotton already bleached, of the right weight, is selected and checked for weaving defects.
2. The fabric is washed to remove any starch (the yarn is starched to strengthen it for weaving).
3. The fabric is thoroughly dried, then cut to size, ironed well and laid out on the wide printing table.

4. Light circular or straight lines, depending on the shape of the tablecloth, are drawn on the fabric to guide the printer who first blocks in the outlines of the motifs.

5. When the outlines are dry he applies the first colour block which should be placed exactly on the outline. Experienced, skilled printers work quickly and accurately. It requires three over-printings, at most four, to complete a motif. An unusually complicated motif may require six over-printings.

6. With the printing completed the cloth is hung above the printing table to be air-dried. Then it is sun-dried and folded away for twenty-four hours.

7. The printing is then checked. If passed, the cloth is dipped in the oxidising solution, mild sulphuric acid, which makes the colours fast and brings them out in a most dramatic way when the cloth is removed into the air.

8. The cloth is rinsed in clean water, dried and checked again.

9. Finally, it goes to the sewing room to be ironed and hemmed.

In the commercial enterprises synthetic dyes are used, and in some places, to speed up production of printing, handscreens are being used and even rotary machines have been installed. However, there is also a move back to traditional methods, to preserve old crafts, with centres being developed where block-printing in old designs is being done using natural dyes based on madder and indigo and producing beautiful rich colours.

Notes

1 Dr Lotika Varadarajan, an eminent Indian textile scholar, has commented on this statement. She thinks the craftsmen probably worked out the dyeing methods for themselves, as caste would have prevented an interchange of ideas. Alchemists, she states, would have been brahmans, while craftsmen were sudras. Also, as Hindu males always wore white she doubts whether male brahman alchemists would have been interested in inventing colours. Women and kings had greater freedom so the craftsmen would have endeavoured to please them. Artists, as separate from craftsmen dyers, were not involved in textile design, though may have been consulted if a design was for the king.

2 'French Documents on Indian Cotton Painting', Appendix A, P.R. Schwartz, *Journal of Indian Textile History*, III, Ahmedabad, Calico Museum of Textiles, 1957, 24-32.

3 The myrabolans, or cadou (genus *Terminalia*), which contains tannin, acted as a mordant; with iron it gave black, with alum red. Mordant is a French word meaning 'biting'. The mordants caused the colours to bite into the cloth, in effect bringing about a chemical reaction which made the dyes fast. The fatty buffalo's milk mixed in solution with the myrabolans acted like a gum and stopped the colours spreading in the cloth.

4 Producing a good black had always eluded the European dyers, so this information which was given in detail in the letter was immensely valuable.

5 *Natural Dyes of India*, All-India Handicrafts Board, Bangalore, 1975; and Rangeen, *Natural Dyes of Bangladesh*, Sayyada R. Ghuznaui, Dhaka, 1987.

6 Following is a technical explanation of the extraction of the colouring matter *indicum*: By fermentation *indigotan* is separated from *indiglucin*. Continued fermentation results in the further separation of indigo white from *indigotan*. Indigo white provides the dye substance which will combine with the cellulose cotton fibre. On oxidation, indigo white is converted into granules of indigo blue. The cakes sold commercially and exported consisted of coagulated granules of indigo blue. *South Indian Traditions of Kalamkari*, Lotika Varadarajan, Bombay, 1982.

6. Design: Paisley, Ikat (tie-dye), Textures

In India, design, like colour, to which it is so closely linked, varies from region to region, reflects nature and climate, and is influenced by caste customs and religious symbolism. As with colour, fashion was never a consideration. Artist-dyers were inspired by religion to produce temple cloths and hangings of the utmost perfection; only the best was good enough for the temple and this attitude influenced all aspects of their work. The quality survives today in designs that can be realistic or stylised featuring human forms, animals, birds, flowers, fruits and leaves as well as geometrical patterns.

Flowers always featured strongly, among the most popular being the lotus and the water lily which, with their delicate colours, voluptuous shapes and large bluish-green leaves, provided a basis for many designs. Another favourite was the hibiscus, a flower with religious significance, used in temple cloths. Several tree flowers and fruits were popular such as golden champa *(Michelia champaca)*, pomegranate with its lovely gold/coral shades, mango, both the deep pink blossom and pink/orange fruit, and perfumed albizia which looks rather like wattle. Other Indian flowers were chrysanthemum, tuberose, and marigold. During the Mughul years many new colder climate flowers were introduced into design, ones that the Mughuls, great garden lovers, had known in northern countries and which grew well in Kashmir; there was the rose, iris, crocus, tulip, carnation and large red opium poppy. These were combined with other Indian flowers to produce marvellous floral designs often in massed effects. With the newly introduced flowers came a whole new range of delicate colours such as pale apricot, translucent green, old rose, mauve, soft yellows and creams. The master dyers spared no efforts to produce shades that matched the flowers and leaves.

Another Mughul design influence was the introduction of a repetitive stylised floral motif called the *buta*. The motif lends itself particularly well to block-printing which features decorative and formalised borders where stripes are used to enclose geometrical forms, while inside the borders are single *butas* or elaborate floral 'meanders'. These designs are seen today in Indian block-printed fabrics from Gujarat, Jaipur and Delhi. They were also a feature of the Coromandel Coast. Reproductions of these designs are seen in modern French Provincial prints.

Throughout the ages trade has influenced Indian design with foreign market demands bringing about changes to the traditional Indian patterns. Such cross-cultural influences are especially obvious in chintz, but most cottons and silks exported from India were adjusted to some extent, as they still are today, to suit colour or design tastes or fashions of different countries. There is still an emphasis on what will sell in the West.

Paisley

An Indian design greatly influenced by cross-pollination and Western fashion tastes was the one we call 'Paisley', probably the most ubiquitous of all designs to come from India, never out of date, always satisfying. It first went to England in the 18th century woven into exquisite Kashmir shawls, and has become a favourite design all over the world. Though it is now used on all types of fabrics from costliest silks and fine woollens to cheapest cottons, in a textile historical context it cannot be separated from the shawls. If ever Indian designers displayed genius it was in this fascinating cone design which has appealed to people from ancient times. It is a fertility symbol that probably came with the Aryans to India but dates back to Babylonian civilisation where it was based on the growing shoot of the date palm. In India the cone shape is variously known as *kairy* (mango), *buta* (floral form) and in Kashmir, where it has become a specialty and where the shawls were woven, the design is called *kalanga* or *kalga* identifying it with the tender topmost shoot of the local cypress tree as it bends in the wind. All these forms usually bend left, seldom right. The design got the name 'Paisley' from the Scottish town where copies of the original shawls were successfully made in the 19th century to meet the huge fashion demand. 'Paisley' has become the generic term, but the French also call it *le dessin cachemire*, and another French term is *palmette*.

Kashmir shawls were first taken to England by the East India Company in the 1760s but did not become established fashion until the end of the century. Napoleon was responsible for taking the shawls to France. He and his officers had acquired them during the Egyptian campaign in 1798 bringing them back as gifts. In France they were at first regarded as male attire, but Empress Josephine made them fashionable for women. She is said to have owned more than 60 shawls and started a fashion for cutting them up into gowns and bedcovers!

Silky Shawls

The intriguing Paisley pattern was new to Europe and so was the idea of a shawl. Before that women had worn a cloak or long cape. These new soft shawls, which could be drawn in close around the body, were silken and warm. 'Shawl' was a new word for the language. It comes from Persian *shal*, a length of fabric which in Persia men wore around the waist. In Kashmir where the winters are cold both men and women wore woollen shawls wrapped around the body. The wool used for Kashmir shawls is called *pashm* and the basic cloth woven from it is *pashmina*. The patterned shawls are called *kani pashmina*, *kani* referring to the wooden spools or bobbins around which the various coloured yarns used in the weaving are wound. *Kani* is also the term applied to the special double interlocked twill technique used to weave Kashmir shawls. The best *pashm* came from the fleece of Central Asian goats that roamed in Kashmir, Ladakh and Tibet, and the ultimate quality, *Asali Tus* came from the underbelly of wild goats, found on rocks and shrubs where it had been rubbed off by the animals as they shed their winter coats. This was the silkiest wool imaginable and was used in weaving the legendary *Shah Tus*, or ring shawl, so fine that it could be drawn through a ring. These special shawls, treasured by Mughul emperors, were feather-light and deliciously warm. Centuries earlier Roman women had bought them eagerly, regarding them as modern women may regard mink coats.

Until the 15th century Kashmir shawls were unpatterned, woven in plain natural off-white pashm, the best quality, or dyed in plain colours. Twill weave was used for warmth. It was a Turkish ruler of Kashmir, Zain-ul-Abidan (A.D. 1420–70) who brought weavers from Turkey to introduce the woven design technique of twill-tapestry weaving. Multi-coloured floral *butas* and *kalangas* were developed and rendered in jewel colours of ruby, sapphire, emerald, purple, yellow on white and black. These remained the traditional colours of the shawls. Hindus preferred more subdued colours on a brown background, the colour of a 'dead leaf'. The Mughuls of course, loved the brightly coloured shawls and also had gold and silver threads woven into the designs. They ordered clothing lengths woven and embroidered, known as *jamawar*, a Persian word, literally *jama* = costume, *war* = yardage. Our word pyjama is similar, *pai* and *jama* literally 'leg clothing'.

When Kashmir shawls first arrived in England all who saw them and felt their exquisite silkiness wanted to own one. It was not long before they became a fashion rage just as chintz had been a century earlier. Genuine Kashmir shawls which were works of art and sometimes took up to 18 months to make were too expensive for the majority of women nor could the Kashmiri weavers meet the new demand from Europe so both English and French weavers began manufacturing shawls 'in imitation of the Indian'. In Britain, at both Norwich and Edinburgh in the late 18th century woven shawls were first produced. However, it was Paisley, the Scottish town, that became most widely known for its 'Indian imitations'. After a great deal of experimentation the Paisley weavers were able to make remarkably good copies of original Kashmir shawls. Paisley, which had been a weaving centre since the 15th century, had always been noted for the high standard of its work. Its weavers, who were highly skilled, worked on handlooms and were masters of their craft, but this was to change with the Industrial Revolution. By the late 1820s shawls were at the height of fashion and the increasingly mechanised looms contributed to meet the demand. Both Norwich and Paisley, and Edinburgh to a lesser extent, were supplying the market with thousands of shawls and this continued for the next 40 or so years. No wonder the 'Paisley' pattern became so familiar. It seemed every woman wanted a shawl. They were exported from Britain to America, and also went to Australia later in the century. Many women owned several shawls; different shawls served different purposes. A 'kirking' shawl with white centre and deep pale Paisley end borders was an essential item in a bride's trousseau, and was worn on the first visit to church after marriage. There were shawls for morning, afternoon, and evening wear; and there were black shawls for mourning. With the great rise of the shawl industry Paisley grew from a small rural town in 1800 to a city full of factories and with a vastly changed way of life for its inhabitants by the middle of the century.

French Style

Although the British pioneered the imitation 'Indian shawls' the French were eventually to become leaders in their manufacture and design. They benefited from one event. This was the ban placed by Napoleon on all imported goods in 1806 which meant they had to concentrate on developing their local industry. The French not only learnt to make good copies of the Kashmir pattern but to apply their own particular design style. Later, from about 1818, they were able to take full advantage of the newly developed jacquard loom with an automatic punched card mechanism by which complicated designs could be produced much more quickly and more efficiently than by the old drawloom, and in more colours.[1]

Import restrictions were dropped in France after Waterloo so that the French could once again buy shawls from Kashmir and agents began going back there, influencing the weavers to alter designs to appeal to French taste. This produced some interesting results. The 'Paisley' cone became larger and elegantly elongated with long streamer-like tips that curved down close to the side of the shape; the designs, generally, became more exaggerated. The shawls were richly decorated and became bigger, long enough to be double-folded over the shoulders and to cover a wide crinoline dress to the tip of a train. French colours, such as magenta, clear blues, yellows and oranges on black were introduced and added new chic. Long multi-coloured fringes were another French idea. French shawl manufacturers also tried to make women buy shawls with wholly European floral designs by French artists, but these were never popular. Women went on wanting the Oriental look, so that no matter how 'Frenchified' designs became the Kashmir origins could always be discerned. Well-dressed English women hankered after the French shawls and, although relations between the two countries were strained, went to Paris to buy them. The French dominated the shawl industry both in Kashmir and in Europe all through the century and their designs led the fashion. It got to the point where

Fashionable French-designed Kashmir shawl showing exaggerated Paisley shapes. Picture is from French La Mode Illustrée, *9 June 1857. (Courtesy the Board of Trustees of the Victoria & Albert Museum, London)*

Indian palampore (bedcover), cotton, painted and resist-dyed, Coromandel Coast, bears English East India Company stamp 1760–1780. The top border is missing but the Tree of Life design with exotic fruit and birds is intact. The mound suggests Chinese influence, 'Chinoiserie'. (Collection: Powerhouse Museum, Sydney)

Lustrous handloomed Indian silk dupion in wonderful colours that are bestsellers in Western shops. (Courtesy The Silk Shop, Sydney. Photographer: Andrew Payne)

Above: Block-printed yardage in traditional natural dyes; left to right: small red flower motif (buta) on blue, from Rajasthan, rose design from Madhya Pradesh, and indigo print from near Jaipur, Rajasthan. (Photographer: Andrew Payne)

Right: Rectangular shawl of silk and wool with elaborate end and side designs woven on a jacquard loom at Paisley, c. 1840–1850. (Collection: Powerhouse Museum, Sydney)

Below: Detail of the same design. (Collection: Powerhouse Museum, Sydney.)

Above: Superb patolu, silk marriage sari, from Gujarat, in double ikat technique, with lozenge design of elephants and tigers, and border tumpal (deep triangles), late 19th century. (Collection Nomadic Rug Traders, Sydney)

Left, from top: Stylised Paisley design on fabric for modern interior decorating designed by John Stefanidis for his current collection of furnishing fabrics. (John Stefanidis ©)

Block-printed square shawl in good quality silk/wool probably made at Norwich, c. 1850. The all-over Paisley design was most popular. (Courtesy Janet Niven Antiques, Woollahra, NSW. Photographer: Andrew Payne)

Ikat handloomed yardage from Andhra Pradesh where the tie-dye technique is known as bandh. (Photographer: Andrew Payne) manufacturers in the town of Paisley were copying French-made 'cashmeres'.

Made in Kashmir

In spite of the good copies that were being made real Kashmirs were always coveted; nothing could match the delicious silkiness and quality of the Himalayan wool. Queen Victoria loved them, wore them herself and gave them as wedding presents. The European shawlmakers imported a certain amount of real *pashm* but supplies were limited. They even tried to import the actual animals; several attempts were made but the goats did not survive either the change of climate or the rigours of travel. The French came up with a cross between the hair of Russian and Angora goats (mohair) which gave a good imitation of the original *pashm*. Then mohair mixed with silk seemed to solve the problem, but this became unpopular when male escorts complained about the fibres rubbing off on their dark suits. Finally, the industry seems to have settled for a mixture of finely spun sheep's wool and silk.

Genuine Kashmir shawls, a few of which are still made, are works of art. The design is worked in the loom by a tapestry twill weaving process using tiny spools *(kanis)* for each colour. Across the weft many colour changes are made with the design being followed rather like a complicated modern knitting pattern. In Mughul times when the shawls reached the zenith of perfection up to 300 colours were used in one shawl which could take up to three years to weave. During the 19th century the number of colours fell to around 60, but it was still slow work. At the fastest pace only a few centimetres were done per day. Traditionally women have sorted and spun the wool, men have done the dyeing and weaving. The main colours and dyes used were deep crimson from cochineal, other reds from logwood, blues and purples from indigo, yellows from saffron, black from iron, and in the 19th century it was discovered that good fast green could be obtained by boiling billiard table felt!

In England in the 19th century to speed up production for the now large demand, shawls, both at Paisley and Norwich, began to be block-printed on cotton. They had already been printed on silk gauze for evening and summer wear but these new ones were printed on cheap cotton and soon flooded the market. A genuine Kashmir shawl took months to make, even a good woven Paisley one took a week to weave, whereas the printed ones could be run off in minutes. The most time-consuming part was cutting the intricate 'Paisley' wood block. Every popular shawl design was copied in the prints, including the 'kirking plaid'. At a distance these cotton shawls looked much the same as the woven ones. Mass production cheapened the shawl and partly contributed to the end of a fashion. Another reason was the change in shape of women's dresses about 1870 from the crinoline to the bustle for which a tight-waisted jacket was more suitable. The last shawl was made at Paisley in 1876. The shawl had also gone out of fashion in France. This meant a great shift of industry in the European weaving centres. In Kashmir the result was tragic with the weavers once again suffering massive deprivation.

Real old Kashmir shawls are now collectors' items. In 1982 the Victoria and Albert Museum in London bought one at auction from Sotheby's described as 'a very fine early 18th century Kashmiri Pashmina shawl, with bright flower sprays in indigo and red on a cream ground', price £3,800 sterling. British and French imitations of the 19th century are also valuable.

Shawl Revival

Traditional shawl weaving by *kani* technique was revived in Kashmir a few years ago with a few master weavers carrying on the old craft. They were helped to survive by government-sponsored handloom development projects. Weaving the complicated cone patterns is still extremely slow with about 1.2 cm (1/2 inch) being woven per day in a 122 cm (48 inch) wide shawl. Even speeding up the work to the limit they could not produce more than 5 cm (2 inches) per day. Embroidered shawls have also been produced in the same designs in beautiful glowing colours on to plain finely twill-woven cloth. The precious goat *pashm* which is now in short supply, is mixed with Australian merino wool, local sheep wool or silk.

Shawls in the same traditional designs are also being made on a commercial basis on jacquard handlooms. These are being produced in various places in the Punjab. Even though the jacquard system cuts weaving time, putting a new design on to the jacquard cards can take up to five or six months, so complicated is the cone pattern. The shawls are one metre wide by 2 metres long. Lengths of up to 20 metres are also being woven for the Western decorating market. The designers usually come from shawl families and, working with overseas buyers, combine traditional designs with new ideas and colours that suit modern taste. Finest wool is used producing twill-woven cloth of high quality though not quite in the same class as the ultra-silken *Shah Tus* shawls woven from rare *pashm* which in India, if they can be found, cost a fortune.

After the shawl fashion waned in the 19th century the cone pattern survived to be used widely on printed textiles. One who promoted it was Arthur Liberty who had opened his shop in Regent Street, London, in 1875, specialising in Oriental art goods. From India he imported muslins, light cottons, cashmere woollens and tussar silk which he had block-printed in reproductions of Indian designs such as the massed florals and the Paisley pattern, all of which are still regarded as 'Liberty' prints.

It is quite a game 'Spot the Paisley' — just look around wherever you go in the world and you are likely to see it on all types of textiles and decorative surfaces.

Ikat

Almost as well-known as Paisley are the designs which result from the complicated tie-dye technique known as Ikat. These designs are visually satisfying and, like Paisley, have been much imitated. They are recognisable by drawn-out, misty-edged arrows, deep flame shapes, diamonds, variegated stripes and zig-zags, in many colours. Ikat-look has become fashionable in interior decorating and is seen in sophisticated machine-made fabrics of all types in jacquard weaves and prints; but these copies, though appealing, have a rigid look in comparison with the beauty and infinite variety inherent in traditional handwoven, hand tie-dyed Ikats.

There is no certainty where Ikat originated, though it does seem to go back in India a long way because Ikat designs are discernible in the Ajanta Cave mural paintings which date from the 2nd century B.C. Some scholars think that the technique may have come from China or Turkestan moving through trade to northern India whence it spread, again through trade, to Southeast Asia. It seems to have gone to Indonesia about the 6th century A.D. although it is thought that the technique may have been evolving there at a much earlier time as it is such an established craft. It was also known across the world in Peru where Ikat cloths have been found in ancient pre-Columbian grave sites.

The word *Ikat* (pronounced 'eekat') comes from the Malay-Indonesian word *mengikat*, meaning to bind, tie or wind around, and this has now become the generic term. The technique is a slow and painstaking one of dyeing the yarn to a pattern drawn in charcoal on the stretched out yarn before it is set up on the loom and woven. This is achieved by wrapping areas of yarn not to be dyed with dye-resistant waxed cotton ties or impermeable leaves. For this operation yarns of required length are tied in bundles or strung on racks as they would be on the loom, then the whole dipped in a dye bath. Tying and dyeing go on until the whole pattern is completed. Care must be taken to ensure that the dyes stay within the pattern; it is difficult to avoid seepage of the dye along the threads, especially with thicker yarns such as heavy cotton, but at the same time it is this seepage that gives some Ikat designs a soft, misty-edged look that is so characteristic and attractive. When the tying and dyeing are completed the yarn is set up on the loom. When only the warp (down) threads are dyed to pattern, or only the weft (cross) threads, this is known as single Ikat. If both warp and weft threads are tie-dyed and matched in weaving, resulting in greater intensity of colour, it is known as double Ikat.

Double Ikats are so treasured that they are regarded almost as religious symbols. The supreme example of double Ikat technique is the silk *patola* of Gujarat. A patola is the ultimate achievement of the tie-dyer's skill and considered one of the most remarkable examples of traditional craftsmanship in the world. Patolas originated in one town, Patan, about the 11th century A.D. among a caste of Jain weavers called Salvi, and they are still woven there by two surviving Salvi families. The craft was revived a couple of generations back after a big lapse. A patola, usually a sari length, is always made of silk. (Strictly speaking the singular is *patolu*, plural *patola*, but they are generally referred to as patola and patolas.)

The word *patola* is derived not from Patan, as might be supposed, but from the Sanskrit word *patta*, meaning silk. The Salvi has an important assistant, the *pattibandha* who is in charge of the tying, and they work closely together. Red, the predominant colour in patolas, is the first dye into which the yarn is dipped; excess dye is removed by washing and boiling, then the colour is fixed by steaming after which the hanks are washed in cold clean water (well water is prescribed).

The process is repeated in further colour sequence, blue, black, green, and yellow. Natural dyes were once used but now have been replaced by high quality chemical dyes. As the weaving proceeds the cloth is closely examined for pattern mistakes and threads replaced if necessary. Traditional patola designs are diamond, circular and lozenge shapes containing floral motifs, dancing girls, elephants, tigers, fish and birds, with *tumpal* (deep triangles) in the borders. Gujarati patolas are dyed and woven with such absolute precision that the warp and weft patterns match perfectly. There is minimal dye seepage beyond the pattern boundaries, no fuzzing or tailing off, resulting in a look of astonishing clarity. Patolas generally take about three weeks to make, but an extremely complicated and large one could take eighteen months. Consequently, the finished article is expensive and considered a great treasure to be preserved and passed on as an heirloom. In the 17th century patolas varied in price from 8 to 40 rupees, a high price then. Today a perfect silk patola would cost from 6,000 to 10,000 rupees, about A$600 to A$1,000. Antique patolas fetch high prices at auction.

Magic Qualities

A patola is regarded as having magical and beneficial qualities, bringing good fortune. It plays an important part in ceremonies, especially weddings, being worn by the bride's mother, and by the bridegroom as a shoulder cloth. It is also traditional custom for an expectant mother to wear a patola sari or to sit on one in the seventh month of pregnancy.

From the 1600s European traders, Dutch and English, exported Gujarati patolas to Indonesia to barter for spices; they were also taken to the Philippines and Borneo. The Indonesians regarded them as exceedingly valuable using them as the cloths of ritual and royalty. Only the wealthy and nobility could afford to own them; anyone who possessed a pile of patolas was rich indeed. The Indonesian weavers had practised Ikat technique but theirs had been single Ikat, both warp and weft; double Ikat was probably not attempted until they saw the Gujarati patolas. There is now one double Ikat, the *Geringsing* produced in Bali, which is so prized that it is considered sacred, mystical and magical. Even today villagers are reluctant to talk about the symbolism of the *Geringsing* for fear of offending the spirits.

Double Ikats are woven in Sarawak and Borneo (East Malaysia). Their designs are inspired by religious symbolism, social customs and everyday events. Old Ikats have pictures of dragons and fish showing a Chinese influence; today aeroplanes sometimes appear in designs. Ikat blankets and hangings are part of the furnishings of the Dyak long houses in Borneo. Beautiful Ikats are woven in Thailand where they have been organised on a commercial basis in recent years, influenced by foreign designers and promoted internationally. The technique has also been revived on the village level both in Thailand and Laos. Ikat technique is practised in Japan where it is known as *kasuri*. The Japanese use mainly natural indigo, and in patterning do not aim for precision, preferring and actually striving for a feathered effect. Ikats were first woven in Europe in the 15th century, at Lyon, France.

In India today single and double Ikat techniques are practised in Orissa state and Andhra Pradesh. In Orissa the technique is known as *bandh*, a Sanskrit word meaning 'to tie', and the cloth produced is *bandha*. In Andhra Pradesh, the technique is known as *pagdu* or *bandhu*. It is in Andhra Pradesh, near Hyderabad, that modern Ikat 'yardage' co-operatives have been set up on a cottage industry basis making warp Ikats in cotton for the export trade. Patterned yarns for the warp are prepared by dyers in 20-metre lengths and modern chemical dyes are used. As of old, these beautiful patterned fabrics are exported to markets around the world and are seen in glossy magazines enhancing the homes of the rich and famous.

Bandhani

Another tie-dye technique that has a long link with the West is Bandhani. It differs from Ikat in that undyed yarn is first woven into cloth which is then resist-dyed by a knotting and stitching process resulting in dotted pattern effects. Our word 'bandanna' for a tie-dyed square cotton scarf comes from the term. Red and white spotted cotton bandannas were exported from India to Africa and the American slave states in the early 19th century and also became popular in England as working men's scarves. In the 1960s Bandhani-patterned Indian dresses became a vogue in Western society. The technique is a specialty of Gujarat; but is also practised in Rajasthan where it is known as *chundari*, and in Indonesia where it is *plangi*.

Textured Designs

Among the most widely used of Indian cottons have been those with subtle textured designs that are woven in the loom. In this category are a wide variety of cottons and cotton/silk mixtures that developed over the centuries through trade and demands of foreign markets. Vast quantities, mainly used for clothing and light curtaining, were imported to England by the East India Company for almost 200 years until the early 19th century; they went to America, and came to Australia. After the Industrial Revolution Lancashire copied them, and they went on being sold throughout the Western world. It is no wonder that many have survived in our own everyday fabrics with names almost unchanged. **Gingham** is one of them. It was originally developed for the ancient Malaysian/Indonesian market and was among those fabrics that the Dutch and English had to secure in order to trade for spices. When later taken to Europe by the East Indian Companies it became very popular.

The name gingham derives from the Malay *ginggang* meaning 'striped' but the fabric was produced in Bengal. Originally ginghams were of mixed cotton and tussar silk, woven with double-threaded warps and wefts giving a distinctive texture, and were made in both checks and stripes. Ginghams of this type are still woven in India and exported through the handloom industries. From the 17th century ginghams became popular in England as an East India Company order dated 1670 shows: 'Ordered from Bengal: 2,000 pieces of striped ginghams according to pattern now sent. 10,000 coloured ginghams, 10 yards long, full yard wide, most Graies [undyed] even colours, free from Rowes [banding] and of best sorts.' ('Banding' is still a problem with Indian handwoven cottons; it occurs because the yarns are of mixed qualities.)

Ginghams were also woven on the Coromandel Coast probably for the Dutch to trade for spices with Indonesia as another English order dated 1689 shows: 'Ginghams white such as the Dutch have of 4 threads, or as we had in Agent Foxcroft's time double-threaded both ways.'

Modern ginghams are sometimes woven with double-threaded warps and wefts, but usually are in plain weave, all-cotton, with squared coloured checks, sometimes narrow stripes, on white.

Other loom-patterned fabrics that were imported from the 17th century and are still well-known to us are:

Seersucker This word is a corruption of the Persian words *shir a shakkar* meaning 'milk and sugar' which aptly describes the practical puckered cloth that we know today. Seersucker was originally woven in silk and cotton stripes, the silk being yellowish, probably tussar. The East India Company, ordering for the London market in 1709, asked for the silk stripe to be in colours: 'Seersuckers, the thick sort 12 yds. × 7/8th yd; instead of being striped with straw-coloured silk, get them striped with pink, blue, gold and other light colours.'

Bird's Eye This design featuring a small dotted diamond pattern is now associated with high quality woollen worsteds but was originally woven in lightweight cotton. The name derives from Hindi *bulbul chasham*, meaning 'bulbul's eye'. An East India Company letter entry of 1662 reads 'We formerly received from your parts (Gujarat) a Striped or Chequered sort of callicoes, called "Birds Eyes".'

Samuel Pepys wrote in his diary, 1665: 'My wife very fine in a new yellow birds eye hood, as the fashion is now.'

Dimity The name derives from a Greek word *dismitos* meaning double warp threads, but refers to an Indian cotton cloth woven with raised warp stripes that was traded around the Mediterranean countries from early times. Dimity was later also woven with fancy figures between the stripes. It was popular in England in the 17th and 18th centuries, imported undyed from India and used widely for bed hangings. In modern times dimity has been woven in muslin weight and used for blouses, nightgowns and children's dresses.

Dungaree We know this now as a worker's overall but the name comes from *dungri*, Hindi for a tough twill-woven cotton cloth usually dyed blue with indigo, but sometimes in brown. The Dutch exported dungri to Malaysia in the 17th century and it later went in large quantities to Europe. At Nimes, in the south of France, the fabric was copied in the 18th century. This cloth 'de Nimes' which became shortened to **denim**, also dyed with indigo, is thus a direct descendant of the Indian dungri, now made famous in jeans, the international garment.

Nainsook This fabric, not textured but popular at the same time, was woven of superfine, soft cotton, and regarded as special. The name derives from two Hindi words — *nain* = eye, and *sukh* = delight. This high quality cotton was considered to give 'pleasure to the eye'. It can be bought today as bias binding.

There are many more of these lightweight cottons with names once well-known but which have now dropped out of use.

Since the 1960s there has been a revival of textured cottons mainly in heavier furnishing weights which have been in big demand on the modern Western market (see 'Australian Connection').

Design in Modern India

Trade and overseas fashion tastes continue to influence exported Indian textiles, but gradually India is changing, becoming more independent and self-confident, less manipulated by overseas demands, and there is a strong return by contemporary designers to an appreciation of the Indian heritage. Young designers travel around India taking ideas from carvings on temples and old houses, from old printing blocks, and from textiles and artifacts preserved in museums. The Calico Museum at Ahmedabad in Gujarat State is a particularly good source of textile ideas. This museum was set up in 1949 by members of the Sarabhai family who own the Calico Textile Mill, one of the biggest in this city of textile manufacturing. The Museum was set up originally in a *haveli*, a traditional Gujarati house with upper balconies trimmed with ornamental carved timber, but has now moved to a much larger, specially designed building. In this fascinating museum are displayed rare painted and printed calicoes, chintz and palampores, canopies, Kashmir shawls, silk patolas, brocades and magnificent embroideries, many originally from the collection of Miss Gira Sarabhai. These treasures provide visual feasts for anyone interested in textiles but are particularly instructive for students of design. Also in Ahmedabad is the National Institute of Design which, as well as training students in all modern media, fosters an interest in traditional Indian design, encouraging its incorporation in modern ideas. Many talented designers are emerging who interpret traditional designs in styles and colours that appeal to both Indian and Western tastes.

Note
1 Although the invention of this loom is attributed to Jean Marie Jacquard, the master Lyon weaver, he actually developed and improved upon a similar loom that had been invented by Basile Bouchon in 1725. Jacquard had his loom ready to use in 1804 but the Lyon silk weavers, fearing for their jobs, rioted against it and burnt the loom publicly. When it did come into use later it benefited weaving generally, and the shawl weaving industry in particular.

7. Silk in India

The Chinese first developed the cultivation of mulberry silk, that most luxurious of cloths, circa 2600 B.C. and kept the process a secret for 3000 years; but India also had silk from ancient times, wild silk spun from cocoons that are collected in the jungle where the silk worms feed on leaves of castor oil plant, and trees such as teak, almond and oak. In the *Mahabarata*, the Sanskrit epic poem describing events that took place around 3000 B.C., a king is mentioned as receiving gifts from a feudal prince of 'furs, woollen shawls, and cloths made of wool of sheep and goats, and thread spun by worms . . .'

When a cultivated mulberry cocoon is placed in boiling water a long strong continuous silk filament unravels which after cleaning can be reeled and woven. When wild cocoons are dissolved in boiling water a collection of shorter ends is left which is spun into yarn in the same way as cotton though sometimes cocoons of the Tussar family can be spun in a continuous filament. Wild silk is still produced in various parts of India and is popular in the West both for clothing and furnishings. Because of the worms' rough and assorted diet the wild silk yarn in its natural state produces an attractive slubby textured cloth in yellowish to brown tones like bark, and because of this it was once called 'bark cloth'. The three main varieties of wild silk produced in India are muga *(Antheraea assama)* which when woven produces a quite lustrous cloth with a characteristic yellow colour; eri *(Attacus ricini)*; and best known, tussar from Hindi *tasar (Antheraea mylitta* or *paphia)* which weaves into a cloth with matt sheen in rich cream to dark brown/beige colours. Its matt appearance comes from the tussar fibre being flat. Tussar was used for weaving mixed cotton/silk gingham. Early European merchants called wild silk 'herba' because they thought it came from a plant or herb, the mistake occurring because the wild silk cocoons look like part of the plant from which they hang. The English 1619 definition of tussar was 'Herba — a kind of Bengala stuff of silke grass called tessar . . .' Tussar was produced in Bengal and at Thana, near Bombay, whence it was exported to Sri Lanka, to Persia and to Rome. The Romans bought a great deal of silk from India among many other luxuries.

The Silk Road

Mulberry silk *(Bombyx mori)* did not reach India, or any other country outside China, until around 140 B.C. after the establishment of the Silk Road which opened up communication between East and West. Trade began to develop with silk one of the main commodities coming from China and cotton going from India. The Silk Road, which always sounds so glamorous, was in fact an enormously long and dangerous route taking caravans often years to negotiate with great loss of life and goods. Coming from China they had to cross the Gobi Desert, skirt the dreaded Taklamakan plateau with quicksands that could swallow up whole caravans if they strayed into it, struggle through icy mountain passes and deal with brigands. At every frontier there were taxes, so that silk became extremely expensive and precious by the time it reached its destinations in the West. There were trading posts along the way with Arab traders going east and meeting Chinese traders coming west. The route went through Khotan and at the Hindu Kush branched down to India where goods were distributed to the domestic markets or shipped from ports to Africa, Egypt and Rome. The Romans regarded Chinese mulberry silk as a supremely desirable cloth and it was worn by rich and fashionable citizens, men and women. The overland route continued west to Persia, Mesopotamia and Constantinople, where silk was sold to Greece. The Greeks called silk *seres*, the Greek for China. The Latin was *sericum*, later *selicum* which became *silic*, finally our 'silk'. In modern Italian silk is *seta*, French *soie*. The silken thread links the countries and the ages. In India mulberry silk was called *Cinapatta* or *Chin-sukh* (China silk).

Fabulous Brocades

The silk which first arrived in India from China included yarn and woven cloth which would have been studied with intense interest by Indian weavers who knew only the rough wild silk. This lustrous *Cinapatta* was so expensive that only the richest people could afford it, so weavers often unravelled the cloth and re-wove it mixed with cotton, wild silk, or wool. The silks that arrived from China were in plain weaves, or in beautiful brocades with patterns the Chinese had developed over the centuries. They wove these on extremely complicated drawlooms similar in principle to the ones that the Indian weavers had also been using for centuries to produce their own complex brocades. These were the *kinkhabs* or *zari*-brocades that were woven with superfine real gold and silver threads on a silk base producing the fabled 'cloth of gold' that shimmered with every move. In the *Odyssey* Homer describes Ulysses as wearing rich brocade woven with a hunting scene, which was a specialty of Bengal, and a vest 'fine as a filmy web that dazzled like a cloudless sun'. All sorts of designs were worked into these amazing cloths such as floral butas, flowering plants and trees, floral scrolls, birds, animals and human figures, and in some regions verses from the Koran and Hindu scriptures.

The drawloom or *jhala* on which they were woven remained unchanged from ancient times until comparatively recently when replaced by the jacquard loom. A British district officer visiting a silk weaving centre in Bengal in 1894 wrote a good description of a drawloom being used there: 'In the Naksha [pattern] loom the draw boy sat on top like a puppeteer manipulating a variety of strings to change the spaces or sheds between the warps and enable the pattern to be made. Harness cords are called Nakshas. For a rich design as many as 14 Nakshas were used. The weavers produce patterns showing horses and riders, and *butas* in florals and other designs.' Like so many of the Indian crafts it did not matter how long the work took provided it was as close to perfection as possible. Drawlooms were used in Europe, in England at Paisley and other places, and in France until the development of the jacquard loom.

The Chinese continued to guard the secrets of sericulture exporting only yarn and the woven cloth. Anyone caught smuggling out silk cocoons and mulberry tree seeds was punished by death. According to legend it was a Chinese princess who managed to leave China with both. She went across to marry a prince of Khotan in the 4th century A.D. and, not wanting to be without silk in her new country, concealed the required 'makings' in her elaborate head dress. Before long the knowledge got to India and mulberry silk was soon being cultivated and woven there. The Indians called this silk *Kauseya* as distinct from the woven silk cloth from China, *Cinapatta*. The secret was taken further west about 100 years later after two Persian Buddhist monks disguised as Christian missionaries penetrated into China, managed to spy out the methods of silk-making and escape with a small quantity of cocoons and mulberry seeds hidden in a hollow cane. They returned via the Silk Road to Constantinople after a journey which took some years and no doubt they received good payment for their efforts. Emperor Justinian soon had a monopoly of silk-weaving in the West and charged more for silk than the Chinese, which was a great deal. In the 10th century silk weaving in the three main areas — Byzantium, India and China — was at a height of production and quality that it is thought has never been surpassed. Arab traders later spread the knowledge of silk cultivation to Italy and France, and both countries became leaders, as they still are, in the industry.

Royal Silk

In India, silk was never manufactured on the same scale as cotton. Cotton clothed the masses, but silk was the cloth of kings. The royal courts employed their own weavers who enjoyed special privileges such as exemption from certain taxes, and provision of board and keep, as well as other rewards.

Fashion plate from The New Lady's Magazine *featuring a rather odd outfit described in the caption as: 'A Lady of Paris ... Dressed in a Taffety polence (skirt), trimmed with an Indian border of a spotted Gauze. There is a round Cap fastened with a head band tied in a loose knot.' Published in London, 30 September 1786. (Courtesy the Board of Trustees of the Victoria & Albert Museum, London)*

The Sultans, immediate predecessors of the Great Mughuls, promoted silk and used it lavishly themselves. During the reign of the last Sultan the state silk-weaving workshop employed 4000 weavers. Delhi, where he had his magnificent court, was at this time especially prosperous, being a centre of trading for the whole of northern India, and this situation was to be developed again with the Mughuls.

When the East India Company came on the scene they did not handle brocade silks as 'trading cloths' as they did the cottons. The rich brocades made for princes and nobility were too exotic in design for Western taste, and too expensive and slow in production to be profitable. However, large quantities of plain everyday silks were imported from the late 17th century in response to a growing fashion for finery. These came from Bengal and though not considered up to the quality of French and Italian silks were much cheaper; and, if selected carefully, good 'Bengals' could be found. A popular silk for petticoats and linings was Bengal 'taffatie' (from the Persian *tafta* meaning a 'glossy twist'). Bengal 'taffaties' were in big demand as a 1680s directive from the London directors of the East India Company to the Bengal factors shows: 'push on the making of Taffeties to your utmost quantities procurable.' People, then as now, liked something new, so 'Of all silk wares take it for certain rule that whatever is new gawdy or unusual will always find a good price at our Candle', a reference to London's candle auctions.[1]

Striped and checked silks and silk/cotton mixtures in different colours were new 'wares' that appealed. Among these were ginghams and seersuckers, and many others with old but once familiar names such as *soosey, tuckrees, mandillas, doreas,* all striped or checked; red and white *charkana* (checks), and *nillaes,* from Hindi *nila* (blue), a striped cloth in sky blue, but also dyed in 'hair colour, red and tawny' with black stripes, and occasionally flowered. Striped silks of these types can be seen in many old French and English fashion plates.

In addition to woven silk, the East India Company was importing huge quantities of raw mulberry silk (unwoven in hanks) to England to be woven at Spittalfields and other centres. In 1793 nearly 800,000 lbs (363 tonnes) of raw silk was exported; the maximum quantity ever exported in one year was 1½ million lbs (680 tonnes). When the East Indiamen, the Company's vessels that plied between India and England, unloaded their large cargoes at the East India Docks on the River Thames no wonder the 'scent of the Orient' floated over London. Preparing the raw silk was a job which kept many English workers busy as, before silk can be woven, the sericin or natural gum has to be boiled off and the silk sorted into quantities before being dyed. Mulberry silk accepts dye readily whereas wild silk needs special preparation like cotton. With silk so readily available and inexpensive in Europe many silk types that had previously been woven in India were now being copied. One was foulard, a twill-weave silk (Indian *foulas*) which was now being woven in France. Such was the reputation of the quality of Indian textiles at the time that the French-made foulard was being sold in Paris as of 'Indian manufacture'.

Silk exporting to England declined after the East India Company ceased trading in 1833, and this adversely affected workers in silk weaving as well as those in sericulture in India, as so much of their production had been geared through this outlet to the English market. The situation became even worse when import restrictions were imposed in Britain due to an economic depression in the mid-19th century and the British silk industry, now unable to get sufficient raw silk, was hard hit, with many well-established silk weaving works having to close. Needless to say the Indians suffered yet another blow to trade.

Liberty Designs

Later in the century a specialist trade in tussar silk to England developed. Tussar had been sold for many years and been fashionable for dresses and light furnishings but no satisfactory way had been found of dyeing it. Now an Englishman, Thomas Wardle, who had a dyeworks at Leek, Staffordshire, and did work for many of the Arts and Crafts Movement people like William Morris and Arthur Liberty, was successful in dyeing the tussar. He was a designer himself with a particular interest in Indian designs and dyes, having travelled throughout India and other eastern countries collecting ideas and organising trade. He imported quantities of tussar silk cloth, mostly undyed, but some with woven designs, one with a lotus design becoming a big seller at Liberty's. For dyeing the tussar Wardle experimented with Indian natural dyes finding them to be much more effective than the synthetic dyes then available; he had done a study of Indian traditional dyeing methods and written a treatise 'Monograph on the Tussar' naming 181 Indian dyestuffs. Wardle made tussar marketable and it became fashionable both for clothing and furnishings. For Arthur Liberty's London shop he produced a range of Indian designs in subtle pastels which became known as 'Liberty colours'. Thomas Wardle's wife, Elizabeth, who was a remarkably gifted embroiderer and founder of the Leek Embroidery Society in 1879, did some beautiful work on tussar in designs based on the Indian poppy and others taken from the Ajanta Cave murals. William Morris had a carpet embroidered to his own design on tussar.

Towards the end of the 19th century the Indian silk industry suffered another blow, almost fatal, when a disease called *pebrone* killed the silkworms. The industry limped along but was not revived to any extent until the early 20th century when silk worms from China and Japan were imported to silk weaving centres in Mysore and Kashmir. In 1910 Sir Thomas Wardle, as he had now become, tried to get tussar production going again in Bengal but it soon relapsed. Like the Indian textile industry, generally, local silk production gradually fell into decline until the great post-Independence revival of recent years.

There is now a big production of mulberry silk in Karnataka state with Bangalore one of India's main commercial manufacturing centres. Sericulture is highly organised throughout the whole state supplementing general agriculture. This carries on an old tradition of the area from the days when Tipu Sultan, one of the last important Muslim rulers with his palace at Mysore, gave his patronage to silk production. Manufacture is mainly by power loom in textile factories, but there is a sizeable handloom industry. Raw silk is sold through a marketing organisation to various other silk weaving centres in the states of Gujarat, Andhra Pradesh, Tamil Nadu, and Uttar Pradesh where Varanasi is the main centre. The latter, once called Banaras, the sacred city on the Ganges, has an ancient tradition of superb silk weaving particularly of brocades significant in religious rituals. The weavers today are indiscernible from the originals. Rich silk *zari*-brocades with repeating *butas* woven in gold and silver on silk were are indiscernible from the originals. Rich silk zari-brocades with repeating *butas* woven in gold and silver on silk were made here for the Mughal emperors and nobility who loved to wear these splendid cloths or present them as gifts to visiting dignatories or as rewards to courtiers. In more recent centuries rich brocades have been made for European royalties notably the Russian royal family. They have become priceless cloths of connoisseurs.

Note

1 The 'candle auctions' which were held at the East India Company warehouse attracted big crowds and there was a festive atmosphere with the arrival of new goods. For the auctions an inch of candle was lit and the best bid received just before the candle burnt out was the successful one. Old East India Company auction advertisements list 'Callicoes, Muslins, Ginghams, Chints, Nillaes, Taffeties' and many others. Retailers bid for the goods to sell in their shops which were also popular meeting places.

8. Mughuls and Magical Muslins

The Great Mughuls who ruled in India for more than 200 years before the British arrived were famous for the extravagant splendour of their courts and their gorgeous clothes. Babur, the first Mughul emperor, had invaded from the north coming from his native Turkistan via Kabul in Afghanistan, where he waited for several years before advancing into India which he knew to be a land of riches, worth conquering. Some of its products he was aware of were sugar and spices, exotic fruit, tropical birds and 'fine white cloth' — India's cotton. He may also have heard of the fabled muslin from Dacca in Bengal, that magical gossamer cloth that had been known for ages past, had been loved by the Romans and the Persians, and was later to give so much delight to all the Mughuls and their ladies. Babur pressed successfully into India in 1526 beating off opposition from the Rajput warrior chiefs of Rajasthan, finally settling at Agra where he established his court. One of the first things he did was to make a garden. All the Mughuls loved flowers and nature, and appreciated the arts and crafts. He was also soon to discover the reality of the country's riches, including the existence of a large textile industry spread over northern India but now disorganised by the upheavals of invasion.

State and palace workshops employing thousands of weavers, dyers and other craftsmen had long been established institutions of earlier rulers. Now the Mughuls were to re-organise them on a grand scale. A large part of the work of the best silk and cotton weavers would be to produce luxury clothing for the emperors, their wives and other women in the harem, and their courtiers. All the Mughuls loved dressing up. A contemporary record has them 'attired in rich cloth of gold, cotton and camlets (mixed silk and wool). They all wear turbans on their heads; those turbans are long, like Moorish shirts; drawers, with boots up to the knees . . .' For every day they wore a long shirt-like garment with tightly fitted top and long sleeves with long loose jackets or kaftans in lustrous silks, finest cottons, or cashmeres. Other outfits were fitted cross-over tops worn with long skirts, or with narrow pyjamas or baggy filmy trousers caught with anklets of diamonds or other precious gems; or fitted jackets with and without sleeves in rich brocades worn with short full skirts and coloured hose. They wore patterned sashes, and turbans of plain muslin or woven with gold thread and sparkling with jewels. In full dress they looked magnificent.

The royal ladies in the harem were also arrayed in fine clothes, their traditional outfit being a long chemise with tight sleeves, and over it another looser chemise, in floating silks, finest cottons or woollens. They sometimes wore four chemises, to give a layered effect. The quality and type of fabrics and jewellery worn denoted status from royal ladies down through the ranks. The eunuchs were also splendidly dressed as they were important people in the harem, much esteemed by the women because they amused them, and by the emperor because they protected the women. Unmarried girls wore round caps with tassels, while married women wore high cone-shaped head-dresses from which hung a floating scarf-like veil. When they ventured outside they covered themselves with 'head-to-foot dresses' and were partly veiled. Conventions must have been less strict at times because princesses are known to have gone hunting, riding and hawking wearing tunics and trousers.

The courts of all the Mughul emperors were dazzling but none more so than that of Akbar, considered the Greatest Mughul. Like all the other emperors, although cruel and ruthless, he loved poetry and literature, and appreciated arts and crafts. His greatness, in an historical sense, lay in his political ability. By being conciliatory to the resistant Rajput warriors and by showing extraordinary tolerance towards their Hindu religion and traditions, even going so far as marrying a Hindu princess, Jodh Bai, daughter of the Raja of Amber (now Jaipur), he was able to win their allegiance and stifle opposition. He increased and stabilised the empire that Babur had founded. With this attitude he had Muslim and Hindu craftsmen working together in harmony, getting the utmost in production; he also brought in many Persian craftsmen to work with the local ones. Unlike rulers of old who had left trade to the merchants and concentrated on protecting and ruling the people, Akbar personally controlled trade and enriched himself through it. There was a huge output of woven cotton and silk, part of which was exported, but a large part was absorbed for use at the court. At Fatehpur-Sikri, the red stone capital he built near Agra, there were 100 craft workshops *(kharkhanas)* employing more than 5000 people. They produced crafts of all kinds including painting, calligraphy, metalwork, textiles and clothing. Akbar himself was a talented painter and craftsman and was often seen in the *kharkhanas* working beside the artisans. Textiles were manufactured here in huge quantities as well as at other centres in northern India such as Bengal.

Palace Luxury

Textiles were produced not only for the large variety of elaborate clothing required by the court but for furnishings for the palaces. Today we visit these deserted buildings, thinking them beautiful but coldly austere. Imagine how impressive they must have been with acres of exquisitely woven silk hangings draped like sails, with curtains, huge canopies of satin or brocade hung with gold tassels, and painted screens. There were magnificent patterned carpets on which were placed cotton floor spreads, like picnic cloths, for eating or sitting upon, painted or printed with flowers such as the popular red poppy. These cloths could take months to make but such was the luxury of the court that they were regarded as disposable, used only a couple of times, then discarded as being too worn for the palace. Silken or painted cotton banners which fluttered and caught the sun were made to be hung outside the palace to greet special visitors, or to celebrate significant occasions. There was no end to the textiles that were required by the court.

Then there were the tents. The Mughuls, originally nomadic, travelled often, on battle campaigns, or on hunting expeditions, or visiting their capitals, and when they went they took the whole court with them including the entire harem. Akbar had 5000 in his harem, including 300 wives. The tent expeditions were like moving cities. François Bernier, a French physician at Aurengzab's court, whose book *Travels in the Mughal Empire, 1656–1668* became something of a bestseller in Europe, went on a tent trip which would have been similar to Akbar's, when the emperor visited Lahore and Kashmir, and he described it in detail. He reckoned that about 400,000 people accompanied the king, including 100,000 troops, 6000 porters for the king's tents alone, and 15,000 porters for the whole camp. Tents of all sizes were needed, and most were made of strong cotton, like canvas. One audience tent, the *Am-Khas*, was large enough to hold 10,000 people! There were tents for every type of service such as kitchens and bakeries, and storage, even a tent for sweetmeats, another for brocade vests, and one for gifts the king would make *en tour*. The royal tents were most elaborate with beautiful linings of woven or embroidered flowered silk, or of the finest painted chintz; some were two-storeyed with hinged doors that locked. And there was not only one set of tents but two! They were called *peiche-kanes* — 'houses which precede' — with one constantly in advance to be ready waiting when the emperor arrived. Bernier wrote that in addition to the huge crowd accompanying the emperor almost the entire population of Delhi went too, including all the craftsmen, even the shopkeepers, as all depended on the court for their livelihood and the king would be away at least a couple of years.

In addition to the tents and trappings, the royal wardrobes occupied a large amount of the weavers' time. Akbar was especially clothes-conscious and was always changing his clothes. Pupal Jayakar writes that he had costumes for every occasion, for different times of the day, and for different seasons. Whenever an unusual colour appealed to him from a flower, or a piece of china, anything at all, he would ask the dyers to copy it. They had to invent a whole new dye chemistry to please him. Here are some of the more unusual colours in Akbar's wardrobe listed in the famous Book of Akbar (*A'in-i Akbari*), a detailed biography written by his friend and adviser Abu'l Fazl: grape, honey, hay, musk, almond, blue like Chinaware, cotton-flower, brass, sandalwood, brownish lilac; he wanted the fragile green of cool meadow grass, the elusive colour of mango juice, the mauve of blossoming iris, and an infinite number of subtle variations of basic colours.

Textiles Destroyed

Gigantic stocks were kept so that the emperor's requirements could be met immediately. A measure of the quantity is shown by records of a disastrous fire that occurred in the cloth store *(toshkhana)* at Fatehpur Sikri during Akbar's reign — ten million textiles were lost including gold cloth, satin and brocades, cashmere woollens, and imported cloths such as European velvets and Damask silk. No modern warehouse could afford, or even have the space, to hold so much stock.

Because of the Hindu influences at Akbar's Court traditional Indian clothing began to be adopted. Akbar himself sometimes wore a dhoti, silken of course, or a Hindu shirt. In the harem the Hindu empress and several Hindu princesses who had been 'presented' as wives, introduced the sari, but the high-born Mughul muslim women, especially the older aunts and cousins, fiercely resisted this 'dress of the infidel'.

The court of Akbar's son Emperor Jahangir was no less splendid than his father's, nor were his clothes any less luxurious. Sir Thomas Roe, who was the English ambassador to the Mughul Court, sent by King James I in 1616, described the emperor's outfit: 'He was covered in huge pearls, diamonds and rubies as large as walnuts . . . and wore a coat of cloth of gold without sleeves upon a fine vest as thin as lawn.' With this he wore a *tus* or ring shawl, a waist-belt *(kamarband)* woven with silk and another woven with gold and silver threads. Sir Thomas wrote home that he felt dowdy in comparison and that the English would have to improve their image if they were to win any trading rights. Jahangir's wife, Nur Jahan, was almost as famous as her husband. A woman of outstanding personality and talents, she was articulate and politically shrewd and had a great influence on her husband. She was also artistic and designed textiles, clothes and carpets.

The next emperor, Shah Jahan, left a lasting legacy of celebrated buildings. Known as the Great Builder he constructed the Red Fort at Delhi, Agra Fort and the sublime Taj Mahal as a mausoleum for his wife Mumtaz Mahal whom he loved dearly. He was a one-woman man who never got over her death in childbirth. Although romantically inclined he was also exceedingly ruthless and exploitative. The luxury of his court, like all the Mughul courts, contrasted starkly with the simplicity, indeed near poverty, of life outside the palace walls. The clothing of the masses was of cotton, basic and utilitarian. Men wore turban and loin cloth, women a sari, or just a waistband with long ends, otherwise were naked. In cool weather they wore quilted jackets, checked or flowered, with fabric depending on caste or means. Hindus wore less structured clothes than Muslims, and wealthier Hindus wore silk. A visitor described the dress of a Hindu scholar: 'He wore a white silk scarf tied about his waist and hanging half way down the leg, and another tolerably large scarf of red silk worn as a cloak on his shoulders.'

Brocade for the Emperor

Both emperors who succeeded Akbar continued his attitude of tolerance to Hindus, but the next, Aurengzab, and last of the really Great Mughuls, became in the latter years of his long reign of 49 years a strict Muslim to the point of bigotry, barely tolerating Hindus. His attitude and actions led to political unrest and ultimately to the gradual collapse of the empire. As a strict orthodox Muslim he was not permitted by religion to wear pure silk which ruled out silk brocade, so the weavers created a new fabric, *himroo*, of mixed cotton and silk patterned to look like brocade and woven on a complicated throw-shuttle loom similar to that used for weaving Kashmir shawls, with extra weft threads of coloured silk, and sometimes gold thread, carried along behind the cotton. Because of the extra layer of weft threads the fabric was warm enough for winter clothes. Himroos are woven today on a jacquard powerloom and used in India for furnishings; the Rashtrapati Bhavan in Delhi, the impressive building designed by British architect Sir Edwin Lutyens and formerly the viceroy's residence, now occupied by the president of India, is furnished with himroo curtains woven in Hyderabad. Handloomed himroos are still made by a few craftsmen but their numbers are declining.

Though Aurengzab became austere in his old age, in the earlier years of his reign his lifestyle was as lavish and lascivious as any of his predecessors' with the palace *kharkhanas* producing an abundance of luxury items. François Bernier described the scene: 'In one hall embroiderers are busily employed superintended by a master. In another you see goldsmiths . . . in another, manufacturers of silk, brocade and those fine muslins of which are made turbans, girdles with golden flowers, and drawers, worn by females, so delicately fine as frequently to wear out in one night . . .' This was the famous Dacca muslin which was so incredibly fine that the luxury-loving Romans called it 'woven wind'. In fact, these Indian muslins were so diaphanous that the Emperor Tiberius, not exactly renowned for his own morals, at one stage forbade Roman women from wearing them, considering them indecent. No fabric ever attracted more romantic descriptions. A Persian poet-laureate to one of the Delhi Sultanate courts could hardly contain his prose in describing the muslins: 'they sit lightly on the body as moonlight on the tulip or a dewdrop on the morning rose.' Some poetic names were *Shabnum* — evening dew; *Abkawan* — running water — because if thrown into a stream the muslin became one with the water; and *Sharbati* — sweet as sherbet. Statistics supporting the fineness of the muslin are astonishing. Jean-Baptiste Tavernier, another 17th century French traveller in India, wrote that the yarn from which the muslin was woven was of such extraordinary delicacy that one pound weight could be spun into a 250-mile length! The ultimate quality, the epitome of perfection, was called *Sangati* — 'for presentation', or *Mulmul Khas*.[1] This was the quality that was presented to the Mughuls who appreciated these muslins to the point of reverence. A piece of this 'Royal Muslin', 22 yards × 1 yard (20 m × 90 cm) took six months to make. Some of the lengths had borders of gold and silver, or coloured silks. When the muslin length was completed a horn was sounded. It was then packed ceremoniously into a hollow bamboo tube, 18 inches (46 cm) long by 1 inch (2.5 cm) diameter, beautifully lacquered and gilded, and paraded through the streets of Dacca, held high for everyone to see, to the residence of the nawab who would send it on to the emperor's court at Delhi. A Persian ambassador at Shah Jahan's Court took back to his king, as a present from the emperor, a coconut shell about the size of an ostrich egg, studded with pearls. When opened it was found to contain a muslin turban, 6 yards (5.5 m) long, reported to be so fine that it could hardly be felt.

How was this 'nothing' yarn spun and woven? The cotton used was of the highest long staple quality. An ordinary spindle was too heavy for the thread so a light bamboo 'needle' rod was used with a pellet of clay at the tip. This rested in a small dish or shell containing water to keep the fingers of the spinner moist. Moisture was also needed to strengthen the yarn so spinners worked in the rainy season, or in the early morning before the dew had gone. The yarn was spun by twirling the spindle and controlling the fibres with the fingers. Tremendous sensitivity of touch was needed and the task was done by young women under 30, considered to be the most dextrous. Once spun, the warp yarn was specially treated over eight days by soaking repeatedly in water, reeling, winding, drying, then soaking in a mixture of water and charcoal powder (soot from a cooking pot). Finally, after further rinsing and drying, the yarn was sized with rice paste, then dried and reeled before being sorted for warping. The warp was set up with three grades — the finest on the right side, next finest on the left, and heaviest in the centre. Royal Muslin was all fine. The weft yarn received less treatment, and was done on a daily basis as required. The weaving was done on a four-post horizontal loom by two men. The shuttle was extremely light, and the beam was locked in place by a stick, with the work exceedingly

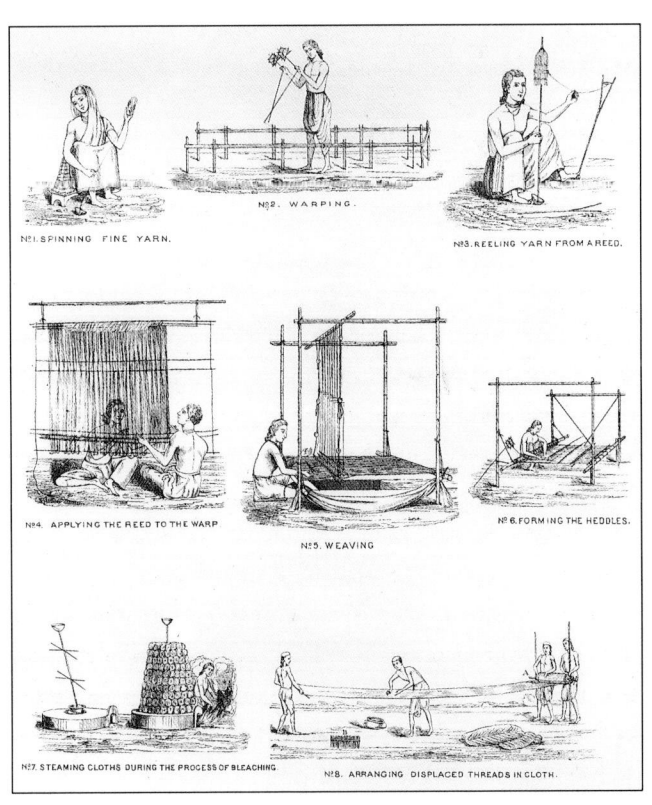

Processes of spining and weaving Dacca Muslin which was so fine that the Romans called it 'woven wind'. It took incredible control to produce this amazing cloth. (From Watson The Textile Manufactures of the people of India, *1866. Courtesy U.S. Library of Congress)*

slow and controlled. When woven the muslin was examined for breaks and repaired by dressers who could replace a thread without it showing, a remarkable feat in itself. It is said that the dressers performed best under the influence of opium which was a common drug of addiction. For taking stains off the muslin they used the juice of amrool plant *(Oxalis corniculata)* greatly diluted. The cloth was given a finish by being sprinkled with rice water while beaten with a smooth chauk shell, then ironed between sheets of paper, lastly packed into a hollow bamboo stick for despatch.

Muslin maintained its high quality throughout the Mughul years but gradually became heavier. For comparison, in 1690 with the English East India Company then trading fairly substantially in Indian textiles, 15 yards (14 m) of muslin, 1 yard (90 cm) wide, was still so fine that it weighed only 900 grains (about 2 oz or 57 g) and cost £40; whereas by 1840 the same measure weighed twice as much and cost £10, meaning that it took a quarter of the time to make, more or less. Superfine Dacca muslins are no longer woven; few people would have the expertise or patience to make the 'woven winds'.

Muslin Craze

Indian muslins of various weights were immensely popular in England especially in the 18th century. Like the chintz craze of a century before there was now a rage for muslin. It is often mentioned in early English novels. In Jane Austen's *Northanger Abbey* written in 1816 there are several mentions of Indian muslin. The hero, Mr Tilney, in early conversation with Catherine, the heroine, prides himself on being a good judge of muslin, and for getting a good bargain: 'My sister has often trusted me in the choice of a gown. I bought one for her the other day . . . I gave but five shillings a yard for it and a true Indian muslin.' Muslins at the time were plain, spotted or sprigged. Another popular muslin-type fabric of the period was jackonet. This word is a French version of the Urdu *jagannath* which was a fine cotton fabric, heavier than muslin, woven at Cuttack, East India, and later developed by the French with checks and stripes.

Another muslin-connected word that came into the language was 'sash' from the Arabic *shash* meaning muslin but which became the generic term for a turban length usually made of fine quality Dacca muslin. Edward Terry, chaplain to Sir Thomas Roe, in describing the dress of Jahangir's nobles, wrote that they wore a turban or sash, 'a long wreath of cloth, ½ yard broad, white or coloured, sometimes interwoven with threads of silver or gold, or coloured silks.'

Imports of muslin from India began to dwindle from about 1817 when Lancashire and centres such as Paisley and Norwich took over, but plain undyed muslin continued to be imported well into the 19th century for printing and dyeing. In recent years plain undyed Indian muslin has had a revival in the West for restoration decorating. Big quantities of plain muslin are also produced for the Indian domestic market where it is used by sari makers who print and dye it to customers' designs.

Cutwork Lace

A fabric related to muslin is *jamdani*, which has geometrical or sprigged motifs on a muslin base giving a pretty lace-like effect. Jamdanis were exported to England from 1670 onwards but only in small quantities because they were described then as 'the most expensive productions of the Dacca looms.' There were also complaints about the quality, according to an East India Company letter dated 1707: 'Jamdanies . . . these sorts of flowered goods if they were made on a fine cloth and not frayey would sell well, but if on a coarse or frayey cloth as they generally come they turn to no account . . .' Jamdani is woven by supplementary weft technique but the extra weft threads are discontinuous being cut to match each motif, unlike himroo where the weft threads are carried along. On the traditional handloom each motif was worked with a separate spool. Today jamdani is woven at Varanasi on jacquard looms in export lengths, in white on white, and like muslin has become popular for curtains in restoration decorating; in fact, with its beautiful lace-like patterns jamdani, or cutwork lace as it is known, is very popular for decorating generally.

Note

1 Mulmul Khas: The Indian name for muslin is *mull* or *mulmul*. *Khas* or *khass* is Arabic-Persian for 'choice' or 'select'. When muslin was exported to Europe from the 17th century the term began to apply to all good, even medium, quality muslin and was corrupted to *cossaes*, *cossar*, and *kassa*. The top quality *Sangati* was not exported to the West, but was reserved for the Mughuls and nobility of Eastern countries.

9. Company and Crown – Into the Raj Years

All through the 19th century the Indian textile industry was in turmoil, tossed about by various events at home and abroad. The most far-reaching and crippling effects were caused by the industrialisation of the textile industry in Britain which began in the 1770s with the establishment of the first Lancashire power-driven mills. Until then the West had been unable to master cotton-spinning but with the invention of the Spinning Jenny a thread could now be produced that was strong enough to be woven into cloth. The first English manufactured cottons were not greatly appealing, and consumers preferred the Indian ones for looks and quality, and because they were cheaper. Lancashire millowners soon overcame this price disadvantage by successfully lobbying the government for a duty on the Indian imports.

After the cotton bans were totally lifted earlier in the century the East India Company had resumed importing Indian cottons and silks of all types in large quantities so that once again they had become part of the English scene with their names absorbed into everyday language. Muslin was in big demand for pretty gowns and furnishings; also popular were the fancyweaves such as ginghams, seersuckers, dimities, striped taffeties and many more, all specialties of Bengal which the English now controlled. Bengal had always been a textile state made fertile by the River Ganges. François Bernier had written earlier, 'There is in Bengale such a quantity of cotton and silk that the Kingdom may be called the common store-house ... not of Hindoustan or the Empire of the Great Mogul only, but of all neighbouring Kingdoms, and even of Europe.'

Indigo was another product imported from Bengal. The goods were carried down the river to the growing port of Calcutta where the East Indiamen, the fine ships of the Company's fleet, waited to take them to London and unload them at the East India Docks.

At Madras the ships picked up more cotton goods, specialties of the Coromandel region, such as beautiful painted and printed chints from Masulipatam, colourful Madras checks, percales, more ginghams, nillaes and big quantities of plain

An East Indiaman, one of the East India Company ships that sailed from India laden with cargo to be discharged at the East India Docks in London from the 17th to early 19th centuries. They were fine ships, built of teak and copper-fastened. As trade with India declined into the 19th century the ships began to be sold like this one, Glenelg, which was under investigation because of a dubious buyer in 1846. (From Illustrated London News, 1846, Vol. 9, in General Reference Library, State Library of New South Wales. G.R.L. ref. F050/129)

undyed calico for the growing textile-printing industries in Europe. Madras had evolved from the original English settlement Fort St George, built in 1640 on a rather unpromising beach-front site chosen by an early East India Company official, Francis Day, because it was close to places where 'excellent long Cloath and better cheape' could be found.

Importing of Indian piece goods, especially plain calico, continued well into the 19th century. The French textile printer Oberkampf, when he became prosperous around the early 1800s, would go to London to buy whole East India Company shipments of calico for printing. He would also buy indigo known as 'English Blue' because of the Company's trade monopoly.

In 1813 the Company lost its monopoly after which trade in India was thrown open to other British companies and to the 'Free Merchants' who had been in India for many years operating under licence to the East India Company but restricted to trading within India and to handling local business such as accounting and land sales. One such 'Free Merchant' was Thomas Parry who had arrived at Madras as a young man in 1788 and was to found Parrys of Madras, a firm which still exists, with its early fortunes based on banking, insurance, shipping and trading in a variety of commodities including indigo, raw cotton and mill-woven piece goods. Parrys rode out Independence to become today one of India's biggest industrial enterprises. An interesting turn of history's wheel is that Parrys, through a subsidiary, is once again exporting handloomed cottons.

As the Lancashire powerlooms became more efficient and the quality of the products improved the demand for Indian cloth declined. English consumers were beginning to accept the good copies that Lancashire could make and duty on imports kept the local prices acceptable. Silk and silk/cotton mixtures that had come from Bengal were being copied at Spittalfields, or at Lyon in France. Muslin was being woven at places like Paisley and Norwich, and Indian embroidered muslins and calicoes were being imitated in Switzerland.

Starving Weavers

By the 1820s Indian imports had fallen off dramatically with weavers once again finding themselves out of work and starving in their thousands. Lord William Bentinck, governor-general

Striped 'taffetie' for dresses like this Paris model of 1827 continued to be fashionable into the 19th century, but, whereas the silk once came from Bengal, with industrialisation it was now being copied at Spittalfields in Englands or Lyon in France. (Courtesy the Board of Trustees of the Victoria & Albert Museum, London)

at the time, and an Englishman who really cared about the Indian people and, in the modern sense, had a social conscience, wrote in a report to Whitehall on this devastation 'The bones of cotton weavers are bleaching the plains of India.' Woven cloth was still required for the domestic market but even this was eroded when British cottons started being exported to India, first spun thread in 1825, followed soon by 'Manchester' piece goods. A duty was placed on these goods

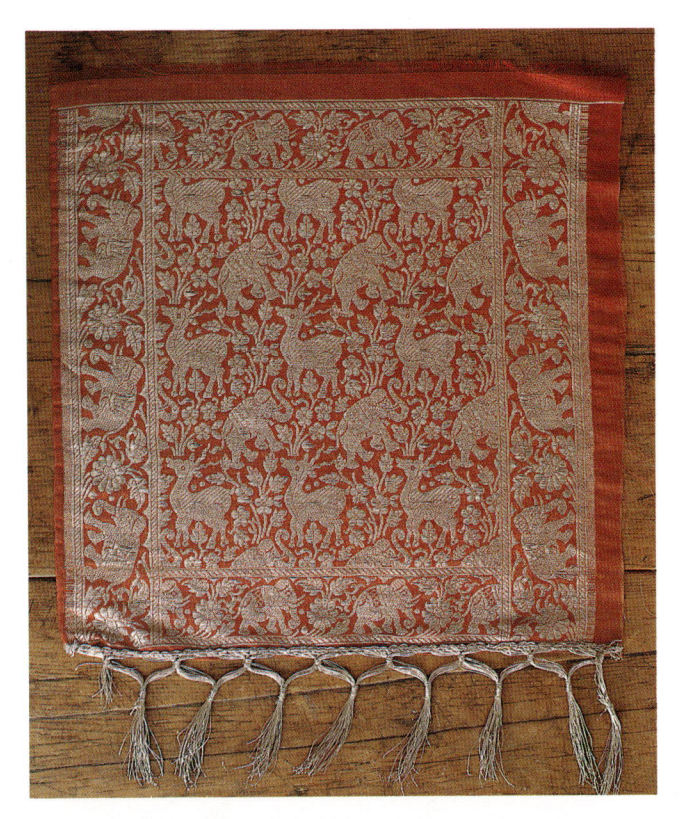

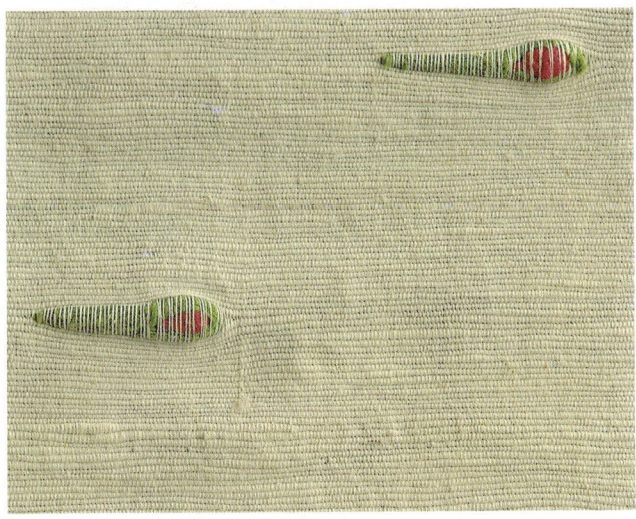

*From top: Brocade featuring a hunting scene (*Shikar gah*) woven recently at Varanasi on a traditional north Indian drawloom, by J. Jaffarali, son of a master craftsman. It is a copy in a generic rather than specific sense of the ancient brocade described as being worn by Ulysses in the Odyssey. It was made as part of a programme of crafts revival, and later the loom was exhibited at the Craft Museum, New Delhi. (Information supplied by Dr Lotika Varadarajan, Indian textile scholar. Photographer: Andrew Elton)*

'Magic Eye', wrapped weft design by Sheila Hicks, with 'eye' highlighted in unusual colours. (Photographer: Andrew Payne)

Right: Patchwork of our everyday fabrics with direct Indian ancestry. From top: denim (dungri), plain dyed cotton, plain and striped seersucker, checked ginghams, raw silk and cotton checks. (Photographer: Andrew Payne)

Above: Religious yellow: saffron-clad pilgrims on the way to a shrine in southern India. Note the touches of dazzling blue. (Photographer: Andrew Payne)

Left: Jayasinhji Jhala, Rajput prince, son of the Maharaja of Dhrangadhara, a state in far Western Gujarat, wears his splendid turban in the state colours of watermelon, navy blue and gold. Each colour has a special historical significance.

Opposite: Emperor Akbar, during whose reign textiles in India reached a zenith of perfection, holds court amidst a dazzling array of colours in the fabrics of his own costume and those of the courtiers, as well as in the canopies and floor coverings of the palace. Picture from a watercolour, copied from a 16th century manuscript 'Anvar-i Suhaili' (The Lights of Canopus) by Baswan, Mughal, Lahore. (Courtesy Bharat Kala Bhavan, Banaras Hindu University, Varanasi)

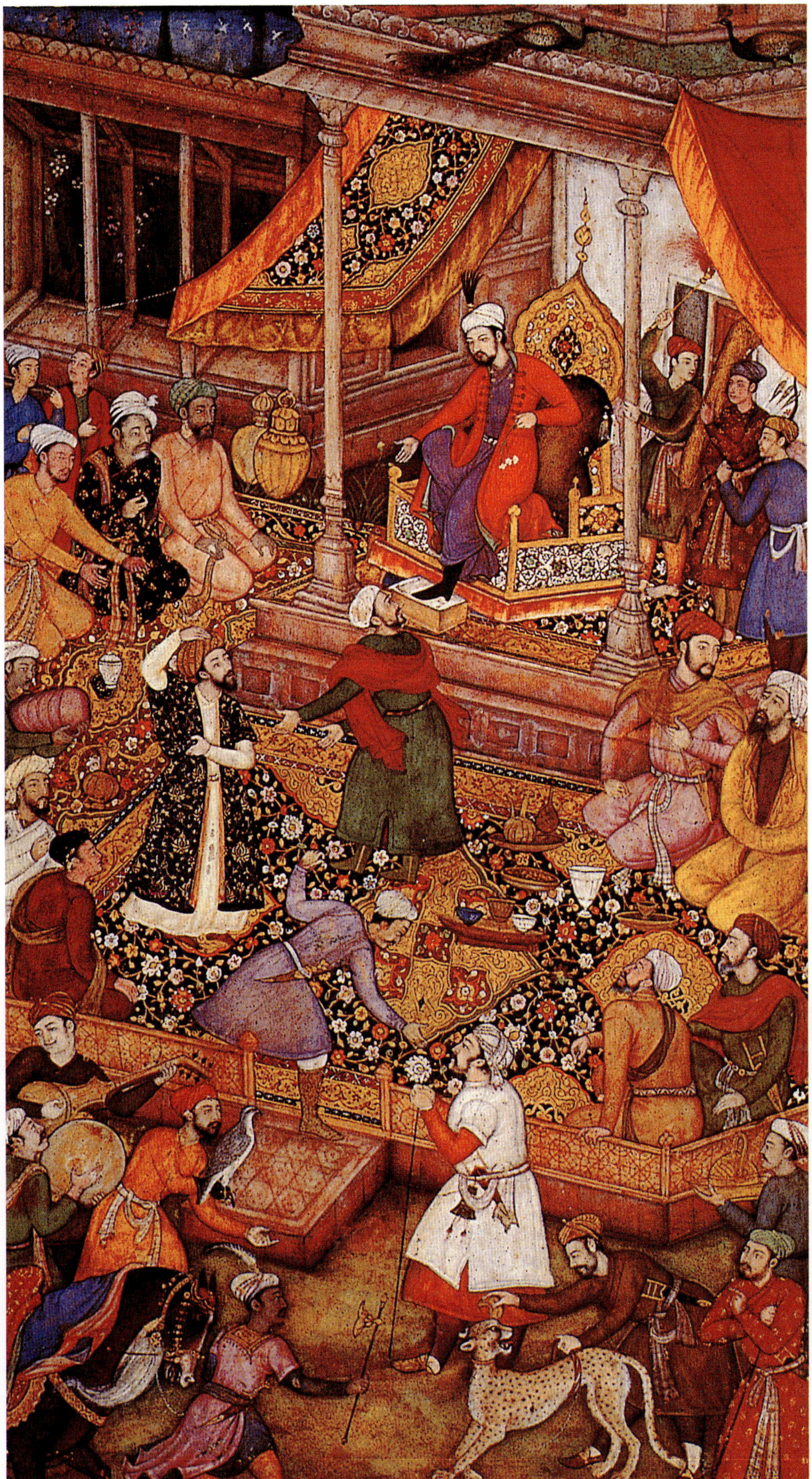

Many early houses in Australia were furnished with 19th century English chintzes. Here in the small sitting room at Vaucluse House, Vaucluse, New South Wales, is a reproduction of an English chintz c. 1830–1840 which matches the period when the room would have been first furnished. (Chintz pattern 'Climbing Geranium', Colefax & Fowler, Ltd ©. Photograph courtesy Historic Houses Trust of N.S.W.)

entering India but on the whole the Lancashire cotton manufacturers were unaffected and unmoved; they were becoming unprecedentedly rich from cotton and Britain was becoming an economically powerful country. But there does not seem to have been much concern for the people who produced the goods. Not only were the Indian weavers disregarded but the Lancashire workers were also shamefully exploited. It is interesting to note an 1833 statistic in this regard which appears in E.J. Hobsbawm's book *Industry and Empire* (Pelican, reprint 1987); while the big mill owners in Lancashire were becoming millionaires the total annual income of 1,778 mill-workers' families, comprising 9,779 individuals, amounted to only £828.19s.7d. The 'shillings and pence' emphasise the paltry amount. In addition there were the iniquities of child-labour, extremely long hours of work, bad conditions and few rights, all of which led in time to revolt and the emergence of the Union Movement.

In India the East India Company continued trading until 1833 then became purely administrative running the country in trust for the British government. There were many problems to deal with, and although they had some outstanding men in their ranks, leaders and scholars who knew Indian languages and understood the people and respected their culture, the social and religious complexities of the country, plus constant difficulties with a few remaining Mughul rulers, made it difficult to govern. Resentment among the people had also been building up for a long time, not only because of trade exploitation but over religious grievances, real and imagined, culminating with the Mutiny in 1857. The following year India was annexed by the Crown and the British Raj years began. John Company's day was over.

In the textile area the main requirement for Britain now was raw materials to feed the mills. At first most of Britain's raw cotton came from the slave plantations of the West Indies, before they switched to producing coffee and sugar, then from the American southern slave states. Cotton from these places was of better quality than the Indian product, longer stapled and cleaner. The invention by Eli Whitney, a mechanically-minded Yankee who had migrated to the Southern States, of a cotton gin for extracting the seeds from raw cotton had been a great advance. Before that it could take a slave a whole day to remove the seeds from one pound (450 g) of cotton. Two events now occurred to upset these sources of supply; the first was a failed harvest in America which led to bad unemployment throughout the industry on both sides of the Atlantic, and the second was the outbreak of the American Civil War which lasted from 1861 to 1865. Manchester, hard hit by both occurrences, was desperate for cotton and turned to India. They also procured cotton from Australia at this time where it was being successfully grown in Queensland.

First Cotton Mills

India was unable at first to produce enough cotton to meet the sudden demand but local growers were soon mobilised to cope with it. It created a boom with attendant problems of inflation, high costs, wages spirals, subsequent busts and bankruptcies. The boost to the economy, however, did result in the building in 1854 of two cotton mills, at Bombay, which laid the foundation of India's modern textile industry. The first mill in India had been built in 1818 in Calcutta. After 1854 other mills followed with ten mills built during the American Civil War period and more later in the century. To work in one of the early Bombay cotton mills Scottish girls, probably trained at Paisley, were brought out, but not being able to stand the climate, the poor girls died off fairly quickly and are buried in unmarked graves in a Bombay cemetery. It seems the ultimate in worker exploitation. Some mills were run by the British, others by Indians, notably Parsees, such as the Tata family who founded the Empress Cotton Mill at Nagpur, Central India, in 1887, and later started the Indian iron and steel industry. Today the Tatas are among India's biggest industrialists with their name on everything from bicycles to trucks, to manufactures of all kinds. The emergence of these capitalists laid the foundation of India's modern middle class. All the mills were to benefit from Lancashire expertise, with British technicians working in or running the mills. These mills produced mainly cotton yarn for which a good market had been found in China and Japan. Lancashire mill owners were so alarmed at the growing competition that once again they lobbied the government for action, resulting in the abolition of tariff on British yarn and woven goods entering India. Throughout the 19th century the Indian textile industry was up against a government policy 'calculated to hamper its growth by introducing tariff legislation which was neither fair nor equitable and which was in the highest degree prejudicial to the interests of India.'

The demand for raw materials in Britain turned India into a country of plantations. Apart from cotton, other textile-related plantation crops were indigo and jute. Jute, which is similar to hemp, is a product of Bengal and was an old cottage industry. It had long been used there for ropes and for *goni* or gunny (sacking). Anyone who has imported Indian handwoven fabrics in recent years will be familiar with the rolls being sewn into gunny bags, as they might have been in the days of early Arab traders. Jute is also often mixed in with cotton in weaving, and gives an attractive finish to the cloth. From 1833 jute began to be produced on British-managed plantations, taken by boat down the Ganges, then shipped to the Scottish town of Dundee which became a world centre of sack and rope-making.

Indigo, next to cotton, was to become the largest and most important of the cultivated crops. Until it was synthesised natural indigo was in demand by dye works all over the world. The East India Company had at first got it from Gujarat, the traditional indigo-producing state, then later, when it became in such big demand in Europe, from the West Indies and North America where it had begun to be cultivated; but in 1779 the Company decided to grow indigo on a plantation basis in India. Indigo production was later developed on a huge scale by British companies and became an extremely profitable export for more than 100 years. This meant that vast areas of arable land were used and many Indians gave up crop farming to grow lucrative indigo, while others were employed on plantations belonging to British companies. With the synthesising of indigo its cultivation dwindled to nothing which meant big losses for the companies and for the Indians partial or full deprivation of livelihood.

Adjusting to the new synthetic or 'tinned' dyes, as they called them, was not easy for the Indian dyers. They found them difficult to handle and their colour scale was crude compared with their own natural dyes. The new dyes were not fast either, and the dyers hated cheating their customers; so they compromised by using natural dyes for quality in conjunction with 'tinned' dyes for speed. One dyer complained that unless he added a natural ingredient to the synthetic dye, particularly red, 'the godown-keeper [warehouseman] returns the cloths which have faded while in stock to be re-dyed.' Synthetic indigo also was found not to be fast so they added lime and treacle and a few other ingredients to make it usable.

There had always been a large domestic market for traditionally made saris but even this was invaded by Lancashire with cheap copies. At first women did not buy these imports preferring the local ones for their artistry and character and for the quality of the natural dyes. The motifs may have been irregular because of the hand-printing but they preferred this look to the precision of the mechanical printing which they found boring. They also preferred the ivory background which came from river washing to the white of the powerloomed cloth. The British manufacturers lost no time in sending out agents to find out why their saris were not selling. The result was they chipped and otherwise damaged their own blocks to copy the handmade look and found a chemical process that could simulate the river-washed background. Soon the Indian buyers could hardly tell the difference and the cheaper copies — 'sinful imitations' — scooped the market. Only conservative older women went on buying the traditional saris. By the end of the century the younger generation even looked down on traditional saris regarding them as having 'clumsy' patterns, in contrast to the 'decent patterns' of the imported ones. Manchester was even copying palampores, those wonderful hangings that for centuries had displayed the ultimate of the artist-dyer's skill. These took months, even years, to make in India; in Manchester, once the block was made, admittedly a skilled and time-consuming job, the cloths could be turned out quickly. They cheapened the superb works of art.

Pirated Designs

The pirating of Indian designs caused great resentment among Indian weavers and dyers. A British official who visited many of the weaving districts in the late 1880s for the purpose of enquiring into the textile industry reported: 'Owing to agents of European firms who have been busy lately buying up native cloths as patterns, the weavers in nearly every place I visited looked upon my enquiries with great suspicion; and in some cases refused to allow me to see their looms.' Many officials sympathised deeply with the Indian craftsmen and deplored the trade exploitation. One wrote, 'The importance of preserving the trades and industries of these clever and patient people cannot be over-estimated.' A district officer, after visiting a weaving town in the Madras area in 1890, reported: 'The town has for ages been the headquarters of a large community of weavers who have of late suffered grievously owing to the competition of Manchester goods.' Similar reports came from other long-established textile areas.

Seeds of Independence

The many grievances in the textile area were adding fuel to the nationalistic feelings that were beginning to be expressed politically through the Indian National Congress. This was an organisation that had been launched in 1885, oddly enough by an Englishman, A.O. Hume, a former civil servant who had stayed on in India. He had formed it with the blessing of the viceroy of the period, Lord Dufferin, who was a liberal-minded man. The aim of Congress was to give Indians more say in public affairs. At first it was a fairly mild secular association of educated Indians, lawyers, teachers and journalists who met annually to debate social reform and report their opinions back to the authorities, but the fact that they met only once a year meant that their 'say' was not great. Nothing much came of it until the 1890s when two outstanding members emerged, G.K. Gokhale, a moderate, and B.G. Tilak, a radical. Between the two of them they moved stronger ideas forward and the 'voice of Nationalism' began to be heard through Congress. Tilak, the extremist, was now saying aloud 'Get rid of the British'.

It would not be long before another voice would be heard, one that would in time become extraordinarily effective, that of M.K. Gandhi. At this time he was still in South Africa where he had set up a successful law practice but had begun devoting more and more time to helping indentured Indian workers who were being badly discriminated against and generally unfairly treated. He had experienced the same treatment himself. By challenging the authorities with great skill and courage he had been able to win many concessions for his fellow countrymen as well as individual legal cases involving injustice. At the same time he became active politically forming his own Congress and publishing a weekly newspaper *Indian Opinion*; he also began developing political behaviourial methods that would later become closely associated with him in the struggle for Indian Independence, notably passive resistance or *Satyagraha*. By the time he returned to India his reputation had preceded him and he was asked to address Congress and speak to large public gatherings in Bombay. What he had to say and the way he said it captured the imagination and people clamoured to hear him. Even at this early stage his genius had been perceived and the honorific title 'Mahatma' meaning 'Great Soul' bestowed upon him. In 1915, during the First World War, he returned to India permanently, and his great work for India began.

10. Gandhi and Threads of Independence

Gandhi promoted spinning and weaving of cotton as a way of Independence *(Swaraj)*. His view was that by spinning cotton and weaving it into basic white cloth known as *khadi*, peasants' cloth, not only would millions of poverty-stricken, unemployed, underfed people have work but be able to clothe themselves; and the general population would also be able to provide its own clothing without having to rely on Manchester imports. He called on everyone regardless of caste and wealth, whether educated or uneducated, to spin a certain amount of cotton every day and wear clothes of homespun cloth. In this way he thought social barriers could be broken down and equality achieved, with people coming to respect manual labour, 'bread labour', as he called it, and weavers and craftspeople being given the status they deserved. He also tried to give dignity to the Untouchables by renaming them *Harajans*, 'Children of God', and worked all his life, unfortunately in vain, for their acceptance in society.

Spinning and weaving became symbols of the National Movement *(Swadeshi)* and khadi became the uniform of Congress and its followers. All disciples of Gandhi had to pledge to wear khadi. The most common outfit was the high-collared jacket worn with dhoti, or trousers, and 'Gandhi' cap. Gandhi himself, who long ago had discarded the black coat and striped trousers of the London-trained lawyer, favoured peasant clothes — dhoti, shawl and sandals. In doing this he felt he could identify with the poorest people. Wherever Gandhi went he dressed in this simple way, even to meetings with the viceroy in Delhi. Winston Churchill was scathing about this: 'It is alarming and also nauseating to see Mr. Gandhi, a seditious Middle Temple lawyer, now posing as a fakir . . . striding half-naked up the steps of the vice-regal palace . . .'

When Gandhi went to London in 1931 for the Round Table Conference to discuss the possibility of Dominion status for India, though it was mid-winter, he wore his usual khadi outfit and certainly did not change when he went to Buckingham Palace to take tea with King George V and Queen Mary. To a friend who asked him if he thought he had been wearing enough for such an occasion, he replied: 'It was quite all right — the King had enough on for both of us.' Though the Palace visit was an ordeal he had managed to keep his rather disarming sense of humour.

Soon after Gandhi returned from South Africa he founded his first ashram, the famous Satyagraha ashram at Ahmedabad. This location was chosen as being the centre of a traditional textile state where a cottage industry of handweaving could most likely be revived. Ahmedabad, which is known today as the 'Manchester of India', already had many big textile power mills. Lacking funds to run the ashram Gandhi had no compunction in seeking help from the successful capitalist millowners. At first they were reluctant to give assistance because they considered that his 'Khadi Movement' which encouraged the wearing of handspun cotton cloth would be competition for their mill-made cloth. However, he convinced them that with so many millions to be clothed all production would be absorbed.

Ghandi's Bungalow

One of the first big owners to come to his aid was a Muslim, Ambalal Sarabhai who owned the Calico Mills. The Sarabhai family, particularly the female members, became strong supporters of Gandhi. In the present generation the family contributes, with other Ahmedabad textile mill owners, to maintaining the ashram buildings in beautiful condition in park-like grounds and has added a Gandhi museum and hostel quarters for anyone who wishes to stay. Gandhi's small bungalow is there with four simple rooms opening on to a terrace overlooking the Sabarmati River. In one of the rooms is preserved his floor-level desk at which he would squat each morning to write lectures or pieces for his weekly newspaper *Young India*, after which he would work at his spinning wheel.

Gandhi and his disciples, the Ashramites, set an example of spinning and weaving. When he first went to the ashram he had never seen a spinning wheel, and scarcely knew what a handloom looked like. Some looms were soon found and the disciples learnt to weave; at first they used mill-spun yarn but the mills, again because of perceived competition, were not keen on supplying this, hence the need for hand-spinning. In any case Gandhi preferred to be independent of the mills. Finding a spinning wheel or *charkha* was difficult. The traditional handloom industry, which had once been so huge, had practically ceased to exist in that part of India. Some old upright wheels were at last located in a dusty corner of a former spinner's house, and a spinner was found who could teach the Ashramites. Gandhi grew very attached to his spinning wheel saying it was an aid to meditation and that it cleansed the soul. He recommended spinning to everyone as a daily discipline. Each morning he would spin for about two hours and often receive important visitors during this period. In his charming biography *Mahatma Gandhi and His Apostles* the Indian writer Ved Mehta gives a description of Gandhi spinning, and although it deals with a much later period, when the Mahatma was quite an old man and living in another ashram, the scene would have been similar:

'His thin slightly nervous hands worked rapidly, the spinning wheel made a soft, warm, comforting buzz, and his lap was soon filled with fluffy cotton fibre. He could spin half a hank of thread, or four hundred and twenty yards at one sitting. This thread ordinarily went into the common pool, but sometimes it was woven into cloth for Ba.'

'Ba', meaning Mother, was the affectionate name given by the Ashramites to Gandhi's wife, Kasturbai, whom he had married when they were both 13. Gandhi's affectionate name was 'Bapu', Father. 'Bapuji' was an even more affectionate version.

Spinning became quite a game in the ashram with all the disciples having to call out how much thread they had spun each day; and spinning and weaving terms became part of everyday speech. They talked of slivers (lengths of cotton fed into the wheel); pirns (weft bobbins); counts (number of yarns

Gandhi spinning with compact improved wheel on his way by ship to the London Round Table Conference in 1931. (Hulton/Deutsch Picture Library)

per unit, for density and weight of fabric); and carding strips (two toothed instruments for untangling or combing the yarn).

Producing cloth for their own clothes gave Gandhi and the disciples a direct experience of a weaver's life and conditions, the limitations of production and the problems of cotton and yarn supply. Gandhi preferred the cloth to be bleached white, one reason being that he had a fetish about cleanliness and this cloth could be washed and washed to look spotlessly clean. Moreover, dyes were difficult and cost more, though some Ashramites did wear khaki khadi, and later one branch of Congress followers, the Pathans, wore red khadi and were known as Red Shirts.

Gandhi used the old-fashioned upright spinning wheel at first until a more flexible, portable flat type was invented. Wherever he went, even to London, he took his spinning wheel and did his daily session. Much later an even more compact wheel was invented by a foreign disciple, Maurice Frydman, a Polish engineer who went to Gandhi's ashram to find spiritual enlightenment and stayed on as spinning adviser. Gandhi was always looking for an easy-to-use efficient and economical spinning wheel that could be used throughout the country.

From the 1920s he began going on tour for several months each year visiting the villages to publicise his 'freedom message' and plan which he called the Constructive Programme. Accompanied by a few disciples he travelled by any means available which included train, car, bullock-cart, elephant and walking long distances. There were reckoned to be 700,000 villages and he hoped to visit as many as possible. He covered a lot of ground because by 1934 there were 5,000 'khadi villages' and by 1940 the number had grown to 15,000. When it is remembered that 'villages' in India can have populations of up to 20,000 and more he reached many people. Wherever he went crowds gathered to greet the Mahatma whom at first they regarded with awe because of his huge reputation, but soon they were drawn closer by his simple message and air of humility. He would talk about the 'khadi franchise', 'yarn currency' and 'thread of destiny'. Gandhi's languages were English and Gujarati, but he soon learnt Hindi and Urdu so that he could communicate with a wider population.

Bonfires

At each village he would inspire the people to start spinning and weaving khadi for their own clothes and to produce a surplus for sale to other villages. Then he would gather the men together and ask them to throw their imported clothes on to a bonfire. It must have been difficult for many to part with treasured possessions but such was Gandhi's persuasive power that shirts, hats, coats and shoes would be tossed on to the flames. Women would also be called upon to do the same with their imported saris, as well as picket and boycott shops and stalls selling foreign-made saris and cloth. Gandhi had started the bonfire idea at the Ahmedabad ashram and even he had been reluctant to throw a beautiful sari belonging to Kasturbai, his wife, into the fire, but because it was made of foreign silk it had to go.

Besides urging the villagers to spin and weave he would tell them to breed cattle so that they would have milk for nourishment, dung for fertiliser and fuel, and bullocks for pulling the ploughs. This was aimed at making them self-supporting. He would also lecture them about cleaning up filth in the villages, impressing on them the need for hygiene to stop the spread of disease. Many Congress members criticised his Constructive Programme as being too simple, too homespun, pressing him to use his energy and talents more in tackling the authorities. While agreeing that it was important to have meetings with powerful people like the viceroy and members of the British government, Gandhi argued that the real strength of the Freedom Movement lay in the millions of people being given a chance to improve their living conditions, which would help them to gain self-respect and sufficient confidence to stand up for their rights. He wanted to rid them of the subservience and fear of authority that he had found so abhorrent.

The Indigo Dispute

There may have been criticism of his Utopian ideas but no one doubted his genius for solving seemingly impossible problems with a combination of a lawyer's skill and unorthodox methods based on truth and morality. As early as 1917 he used *satygraha* (passive resistance) to win an exploitation dispute concerning indigo growers in North Bihar, near Nepal. For years farmers in this area had been cultivating indigo on land rented from British companies. They had been required to grow indigo on a proportion of their land and hand over the annual harvest as part payment of rent. When the planters saw the end of lucrative indigo through synthesising of the dye they said nothing to the ignorant farmers but told them they could now keep the indigo crops and pay extra rent instead, thus hoping to cover a proportion of their own losses. At first the farmers were pleased to accept this new unexpected arrangement, which seemed to them favourable, until a few discovered the truth; but then, when they refused to cooperate, agents of the planters threatened them with reprisals. So they sought help from local lawyers who could do nothing against the powerful planters. Gandhi was then approached with the problem and agreed to visit the area and investigate. It was always his way to find out absolutely every fact before attempting to resolve any problem. Soon after he arrived and began making enquiries he was served with an expulsion order and had to appear in court to defend himself. By a combination of lawyer's skill and knowledge, and his own stayput passive methods, he managed to resist being expelled and stayed on collecting stories from thousands of farmers. Ultimately, he had enough evidence to expose the bullying, unfair tactics of the planters and present a convincing case to the authorities, eliciting from the lieutenant-governor of the state, Sir Edward Gait, the promise of a commission of enquiry. This was set up and found in favour of the farmers. The planters, who were furious, were forced to repay a large proportion of the ill-gotten rent. It was a triumph for Gandhi.

From talking to the farmers Gandhi had been so appalled by their illiteracy and general ignorance that he decided to set up schools in the district teaching, in addition to elementary reading, writing and arithmetic, such basic knowledge as hygiene and good manners. He brought in several disciples and friends to act as teachers. These schools existed for a few years but petered out. Today there is theoretically universal education in India but not nearly enough schools to go around.

A year after the Indigo Affair Gandhi was successful in settling a pay dispute between textile workers and millowners in Ahmedabad. This was a difficult situation for Gandhi as it meant confrontation with owners who had become friends, such as Ambilal Sarabhai. In dealing with the strikers he extracted a promise from them not to use violence and when they broke their promise he staged the first of his famous 'Fasts' in an effort to shame them into returning to passive resistance. The owners thought he was blackmailing them and complained, but, becoming afraid for his health, gave in, offering a compromise on wages which was accepted. Gandhi always encouraged people not to be greedy but to accept a compromise once the principle had been won. Gandhi's methods had worked again and he was a hero all round. Ambilal Sarabhai later confided to Gandhi that he had given in partly because of pressure from his wife and sister who both greatly admired him. Ved Mehta reports in his Gandhi biography a meeting that occurred some years later with Mrs Sarabhai when she told him, 'There was something in Bapuji that was irresistible.' She herself started wearing khadi about 1928 because she believed in its message; here was a woman from the wealthiest class of Indian society who could have afforded clothes from top Paris designers if she had so desired yet was happy to wear the 'freedom cloth'.

Royal Disciple

Gandhi had followers from all classes, from both Nehrus, father and son, and people like the Sarabhais, to the lowliest peasants. One of his followers was the Maharani of Gwalior who as a young student in the thirties had been converted by the freedom message and later as a married woman joined the Congress Party. People like this, representatives of the whole population, were behind him in all his 'Freedom' demonstrations such as the famous Salt March in 1930 held in protest against the salt tax which Indians found especially iniquitous as salt was such a necessity in a hot country like India. The March began in Ahmedabad, led by Gandhi, accompanied by 78 male disciples, bound for Dendi, a beach 338 kilometres south on the west coast, where saltwater evaporated on the mudflats leaving a salt deposit behind. By the time the March reached Dendi it had been joined by thousands from the villages, including hundreds of women, everyone dressed in white khadi. The idea was to gather the crude salt and distribute it free to the population. Gandhi scooped up the first handful, others followed and soon salt was being passed around the whole gathering, and in time around the whole country through a well-organised distribution programme which continued for some time. Naturally, the authorities would not stand for this open defiance of the law and, ultimately, but only after much difficulty, closed the 'free' salt depots and sent the leading dissidents, including Gandhi, to jail. People went on being defiant, even local Indian officials employed by the government. The Salt Affair signalled the start of the Civil

Disobedience campaign that ultimately resulted in close to 100,000 people being put in jail.

Gandhi had several foreign disciples, one of the most devoted being an English woman, Madeleine Slade, daughter of an admiral. She came to visit him at the Ashram and stayed. Gandhi, who changed her name to Mirabehn, became very fond of her, treating her as a daughter. She was devoted to Gandhi to the point of an obsession that in the end so got on his nerves that he sent her away on travelling jobs, and later put her in charge of his press relations. However, she was never far away, always supportive, and even went to jail in the 1930s when there was a round-up of just about everyone prominent in the Freedom Movement. Contemporary newspaper photographs show her as an imposing figure appearing tall beside Gandhi who was 165 centimetres (5 feet 5 inches), and dressed from head to toe in a white khadi sari. After Gandhi's death in 1948 she stayed on in India for several years taking the khadi message to the villages, but in the 1950s left to live in Austria where Ved Mehta met her more than 20 years later and he has described that meeting in his book on Gandhi. She was by this time an elderly woman, with shoulder-length grey hair and walking with a stick. When he asked her about life with Gandhi she was evasive and to the questions 'Do you ever wear khadi or do any spinning?' and 'Are you still a vegetarian?' she replied, 'Those things are not suited to European life.' He found that she had returned to her first love, Beethoven, whose life and music she was studying with the same passionate intensity she had given to Gandhi's cause. She died in July 1982, aged nearly 90.

From 1925 the khadi industry was run by the All-India Spinners' Association which supervised distribution and sales according to Gandhi's original ideas. However, by the mid-thirties khadi began to become more commercially directed, especially in the cities; colours and printed designs were introduced and Gandhi's plan that khadi would provide cheap cloth and labour for the destitute was receding in favour of profits. The 'khadi spirit' was being lost so a renewed moral phase was introduced whereby the khadi supervisor in each village, instead of being an entrepreneur seeking good markets, became more of a social worker dealing with village problems and ensuring weavers and spinners received fair wages.

During the Second World War, because of shortages, khadi was in big demand. In one nine-month period 15,000 'khadi villages' produced 16 million yards of cloth providing work for 3.5 million people. In spite of the demand prices were kept level while mill-woven cloth prices soared. Many more villages would have joined the khadi Movement had not the British Raj stepped in, regarding the production of khadi, the 'Freedom Cloth', as subversive. In many areas the confiscation or burning of stock was ordered with workers restrained and those who resisted sent to jail. The industry suffered a severe setback. In the run-up to Independence khadi had another revival urged on by Gandhi himself. To all freedom followers he appealed, 'Spin, spin, knowing the full implications of spinning. Let all those who spin wear khadi and let no one who wears khadi fail to spin'.

Some of the more sophisticated Congress members, although they wore clothes of khadi, did not really approve of Gandhi's peasant outfit. Jawarlal Nehru, his heir-apparent, was one; he himself always wore a high-necked khadi shirt with trousers, or a beautifully cut cotton or woollen jacket with high collar. This 'Nehru jacket' combined with a 'Gandhi' cap of khadi became his uniform.

Nehru worshipped Gandhi. When the great man was assassinated Nehru, who on Independence had become the

Gandhi with Mirabehn (Madeleine Slade), his English disciple. (Courtesy National Gandhi Museum, New Delhi)

first prime minister, broadcast the news with a choking voice: 'The light has gone out of our lives and there is darkness everywhere . . . Bapu, as we call him, the father of our nation is no more . . .'

Khadi after Gandhi

After Independence the khadi industry was reorganised under the All India Khadi & Village Industries Board and assisted by government subsidies. Khadi has remained associated with the Congress Movement though most Congress people now wear only the symbolic Gandhi cap in khadi.

Over the years levels of production have remained fairly constant with the cloth being distributed to the rural areas through *bhavans* (shops) to the people who really need it. But there has been criticism from the early eighties that the industry has become too commercialised, with an increase in production for the wrong reasons, at least by Gandhi's standards. In fact if he were looking down from above he would not be at all happy with what he could see. Khadi has become a high fashion fabric in demand by the elite in India and for export to Western markets. It has become chic to wear handspun cloth. Fashion garments are being made in India of khadi in different plain colours, also in checks, stripes and blockprints, and exported to boutiques in London, Paris, New York, and Melbourne. Khadi in plain basic white is also being used for curtains by Western decorators. Khadi in cotton comes in various weights from the heavy traditional type, to muslin lightness. Silk is also used for khadi, both mulberry and tussar; and there is now a polyester/cotton khadi which is popular because it makes the cloth easy to wash and dry.

It is thought that the reason why khadi became so popular as a fashion fabric is because of the movie *Gandhi* — an ironical twist, and one which the Mahatma may, or may not, have appreciated.

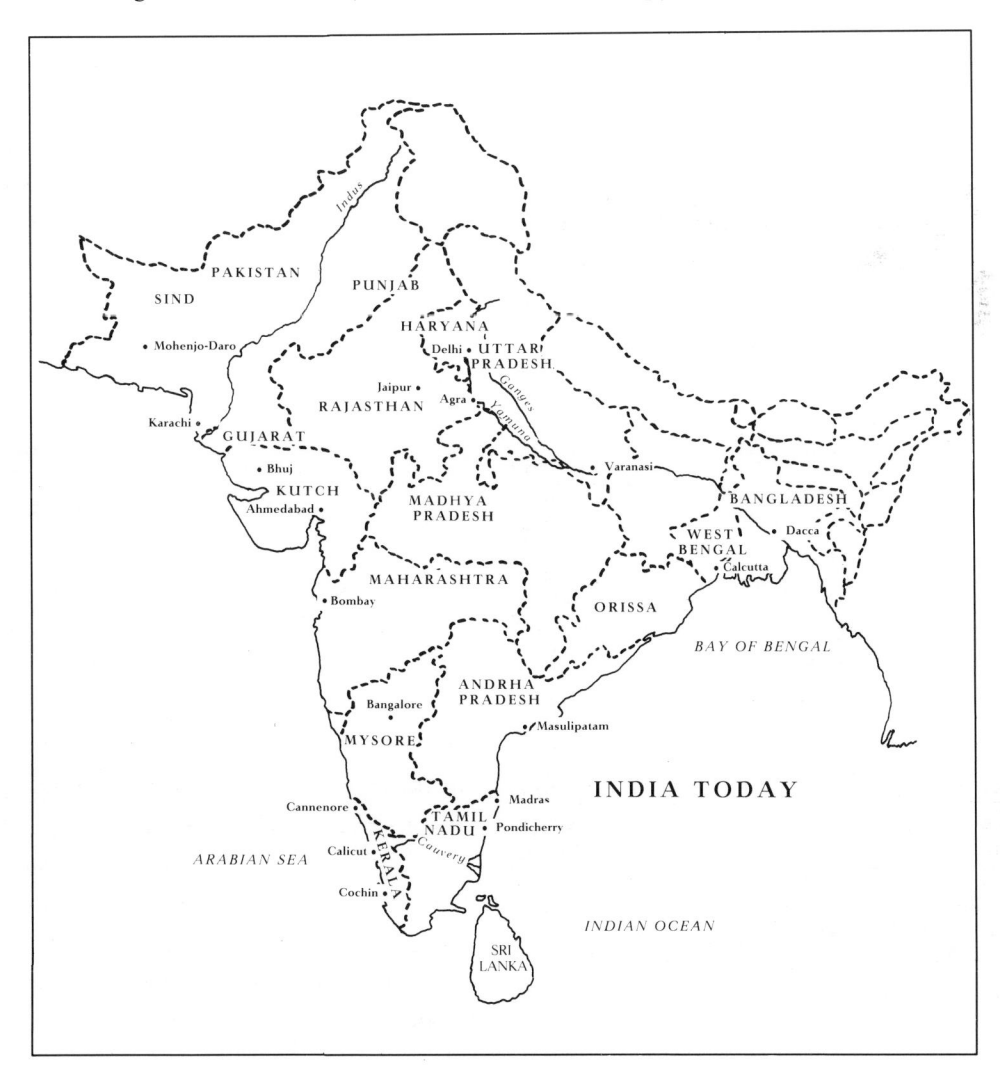

INDIA TODAY

11. Revival of the Handloom

The post-Independence revival of the handloom industry began to gather real momentum from the early 1960s. Behind the movement was the official All-India Handloom organisation supplying dye knowledge and technical assistance to weavers. But it was also greatly aided by several outstanding entrepreneurs or merchants and designers who provided incentives for production. They were people who, far from exploiting the weavers, encouraged and guided them to produce work of quality that would sell not only on the domestic market but in the discriminating West. An outstanding one, and a pioneer of the revival, was John Bissell, an American who, as a textile expert, had gone to India with the Ford Foundation as a consultant to the All-India handloom section. Later he stayed on in India to set up his own firm Fabindia and played a big role in guiding village weavers to produce cloth in colours and designs that would appeal to Western taste. He was at the forefront of handloom exports. He and others who followed him acted like the traders and merchants of old by finding overseas markets and commissioning the weavers to produce cloths suitable for specific markets. It was a continuation of the cross-cultural influences that have affected Indian trade fabric designs from ancient times. The revival was also helped by the growing popularity in the West in the sixties of natural fibres; there was a move away from synthetics to the natural look. The textured village loom cottons from India had immediate appeal to consumers everywhere.

These entrepreneurs gave a fair price for work and assistance with easy loans to buy looms, yarn and dyes. Their policy was to help the weavers help themselves. An example of this concerned a Muslim weaver named Kamalludin who lived with his large family in a village near Delhi in extremely humble circumstances. He had been working for another weaver earning scarcely enough to feed his family but was able to save up to buy sufficient yarn to weave one metre of a fabric he had designed himself to take into Delhi to show an entrepreneur. His talent was immediately recognised and after a few changes to the design he was commissioned to

John Bissell, American-born pioneer of the revival of Indian village handlooms, in his New Delhi store holding a length of 'Rajput' cloth.

weave saleable lengths and lent sufficient money to buy yarn and a loom.

The fabric, a rib-striped cotton called 'Rajput' was extremely successful. It sold well everywhere around the world, used for bedspreads, curtains, chair covers, even for clothing such as skirts and jackets, and goes on selling. Within a few years Kamalludin had his own factory with several looms and weavers working for him. The factory, like all the village structures, is flimsy by Western standards with open sides and mud floors on which women squat to spin, but these shelters are practical in a hot climate and are able to weather the monsoons. Kamalludin, with his success, was able to educate his children and provide well for them.

Into the West

By the early 1970s village loom furnishing fabrics were becoming fashionable on a really wide scale in the West. This great explosion came about largely through the efforts of one particularly energetic and ambitious entrepreneur named Raj Kathuria who took samples to the United States and won orders from some big furniture manufacturers. The result could be seen in cities like New York where the furniture departments of large department stores and small furniture shops along 3rd Avenue were full of sofas covered in thick-textured natural Indian cottons. It set a fashion that went on for several years and spread to many countries, including Australia and New Zealand. In one peak year Raj, the exporter, sent one million metres to various overseas buyers. His biggest single order was for 45,000 metres of a thick fabric called 'Punjab' for an American furniture manufacturer. For handwoven fabric that was a huge order and kept a great many weavers busy. In Australia a fabric called 'Bengal Barsat' with a natural thick oatmeal texture became extremely popular for upholstery and thousands of metres were imported. At Ascraft Fabrics in Sydney so much Bengal Barsat was arriving that extra warehouse space had to be rented to accommodate it all. It was a sign of the popularity of Indian handlooms at that time that the warehouses were soon emptied out, and so it went on until the fashion for it waned.

At the height of the Indian cotton fashion in the mid-seventies to early eighties many competitors came on the scene with a wide variety of types and names. Unfortunately, because of the demand, everyone and anyone in India joined the handloom industry and the quality suffered giving a bad reputation to all Indian cottons which was entirely unjustified. Ascraft's village cottons were always of high standard being quality-checked both in India and in Sydney before being sold to customers. They came from well-established looms with good weavers, but it took a long time for people to realise that every country produces good and bad products, and that it was unfair to dismiss all 'Indian cottons' as inferior. At the time of writing this stigma has not been entirely overcome.

The coloured cottons from the organised and long-established factories in southern India, with their proven high quality of manufacture and reliability of best dyes, have gone a long way to salvaging the reputation. These beautiful cottons, many of which are put together in co-ordinated collections by leading designers and manufactured with strict quality requirements, are used at the top end of the decorating markets. Two people who played an important role in having these fabrics so well accepted were Mr Lionel Paul, overseas sales manager of a hand-weaving factory, The Commonwealth Trust, in Calicut, Kerala state, and Sheila Hicks, the internationally recognised fibre artist whom he persuaded to go there to work with the weavers and help them to produce fabrics that would be saleable on Western markets.

The factory had been producing highly regarded fabrics and had an Indian government contract but when Lionel Paul showed them to English buyers he was told that the colours and designs were 'too Indian'. Hence his approach to Sheila Hicks, who would produce the Kerala Collection.

This factory has an interesting history. Unlike the village looms in northern India it had been operating continuously since it was founded in the mid-19th century; and unlike the flimsy village structures the factory is a solid handsome two-storeyed building with about 150 looms, a dye section supervised by qualified experts, a design department, and a quality-control section. The fabrics are meticulously woven, checked and packed. The enterprise grew out of a German-Swiss Lutheran Mission weaving establishment founded in Kerala in 1844. When the Hindus were converted by the missionaries to Christianity they lost both their caste and their jobs, so the Mission set up trade establishments to employ them. They started a printing works, ceramics factory, and hand-weaving establishment at Calicut, all of which are still in existence and thriving. In 1850 they also set up a textile power mill at Cannenore, a town further up the Malabar Coast.

Lionel Paul, overseas sales manager of the Commonwealth Trust, who introduced high grade southern Indian handlooms to Europe's top markets. (Photograph by author)

Sheila Hicks, American fibre artist, who worked with south Indian weavers to produce the outstanding Kerala Collection of handloomed cottons.

Both textile factories benefited from having German-Swiss dye chemists working with the Indian master dyers. In a quest for more customers at the Calicut factory one of these experts invented a dye colour that would provide camouflage in the jungle hoping to appeal to big game hunters of the British Raj and Indian princes. He called the colour 'khaki', Hindi for 'ashes' or 'dust', and the original khaki cloth produced was a cotton drill. The factory's hope for more customers was more than justified when it received a visit from Lord Roberts, Commander-in-Chief, British Army in India, 1885-93, who decided to take up the cloth for army uniforms which were first worn by British soldiers fighting on the Northwest Frontier. Subsequently, khaki was adopted by the entire British Army. The factory at Cannenore benefited later from technical advice given by weaving experts from Lancashire.

During the First World War both weaving establishments, along with other assets of the German-Swiss Mission, were confiscated as enemy property and put into a trust at first administered by Parrys of Madras, that old East India Company descendant, but later handed over to the Commonwealth Trust which was directly responsible to the viceroy of India. The factories were regarded as important by the authorities being flourishing concerns employing many people and producing goods of high quality. The handwoven fabrics were much in demand by the British ex-patriot population in India. After Independence the various enterprises continued to prosper.

Thread 'Link'

When Sheila Hicks was approached to go and work in Calicut she was relatively unknown, but in the event no one more suitable for the task could have been found as she had already acquired an intense interest in and knowledge of indigenous weaving. In Peru, for instance, techniques had left a lasting influence on her style with the eccentric split-weft weaving and wrapped 'spaghetti' ends inspiring the dramatic wrapped cords and horizontal sculptured effects that have become such outstanding features of her work, and were incorporated into some of the fabrics of the Kerala Collection, such as 'Badagara' with dramatic wrapped wefts and another unusual wrapped weft design called 'Magic Eye'. From Peru Sheila Hicks went to Chile where she organised a group of traditional weavers to produce saleable cloth using local wool, and also held her first solo exhibition. She then moved to Mexico to live and work by which time it was becoming her ambition to 'link the world with a thread'. From Mexico she contributed to an avant garde exhibition at the Museum of Modern Art, New York, and received her first real recognition as an important textile artist. Her next move was to Paris where she set up an atelier which was to become her permanent workshop and where she created her massive wall hangings commissioned by international architects for buildings in France and the United States. Not only is the concept of these various works inspiring but their meticulous craftsmanship is impressive. It was from Paris that she went to Calicut which, at that time, before it got an airport, was an extremely remote place. She was surprised, therefore, to find a factory that was efficiently run and well-equipped, with highly skilled weavers and dyers. They had jacquard looms and the weavers were capable of doing any type of weaving. Her first task was to go through the archives and in due course she developed a set of designs based on the traditional weaves but with some of her own ideas such as the padded wrapped weft effects. Apart from the appealing designs what made the Kerala Collection so outstanding was Sheila's superb colour selection. She paid several visits to Calicut over a few years.

Sheila Hicks has an ability to communicate easily with people of all cultures, and to break down barriers with sensitivity. In Calicut she became a legendary figure to the weavers behaving in the most informal way, wearing a long white cotton shift, identifying with the weavers, sitting down on the ground with them to have lunch. They could not believe that a Westerner could be so friendly.

When samples of the Kerala Collection were taken to Europe they were soon successful and raised the image of Indian handlooms in Western eyes. In Scandinavia, Paris and London they were not only received with great interest but with firm orders. Terence Conran, founder of the UK Habitat shops, became a big buyer for his special Conran shop. The fabrics are sold at Liberty and Harrods in London and are used by many leading interior decorators. In America there are famous buyers such as Brunswig et Fils, China Seas, and Lenor Larsen, who had Commonwealth Trust weave fabrics to his own designs. In Denmark a big home furnishing store commissioned its own collection of line checks and stripes to suit Scandinavian taste but which became popular all round the world. Ascraft Fabrics in Sydney sold all the designs and most colours well for several years, especially the Bamboo range of plain colours, adding colours of their own which became part of the permanent collection.

Another designer who should be mentioned, one who has made a big impact on the look and quality of handlooms, is the Bombay entrepreneur and designer, Shyam Ahuja. He was the first to introduce decorator dhurries in designs based on traditional but with European, such as French, influences, and mostly in pastels on off-white. These dhurries had immediate appeal when they first came on the market and sold well in the United States and Europe, also in Australia. They are woven on flat-bed looms in a controlled situation in Varanasi and Shyam has the best carpet wool used, a combination of Pakistani and New Zealand wool, and best dyes. Unfortunately, although his designs were patented, they were widely copied in India in inferior wools and dyes and so could under-cut the genuine Shyam Ahuja dhurries. Shyam has also produced a co-ordinated pastel range of cottons, and some exquisite pastel silk patolas (ikats) made in Gujarat.

The popularity of village loom fabrics with export markets continues, in fact there has been a recent revival of demand for the naturals; but on the whole they have given way to the high quality factory handlooms in a variety of colours. There is a growing demand for made-up items such as tablecloths and mats, bedspreads and other homewares. All this gives plenty of work for the weavers. By our standards, though, they are not well off. They do mostly piece work so there is no guarantee of continuity. There are no pensions or paid leave, no benefits of any kind. This is not so in socialist Kerala state where conditions for workers are much better. At Commonwealth Trust, for instance, there is holiday pay, meals are subsidised in the canteen, and there is a child-minding creche. The employees are also paid reasonable wages. On the other hand in the Kerala factories workers tend to go on strike, often for extended periods, for better pay and conditions. Cynics could say that this is a result of better all-over eduction in Kerala than in the rest of India. Unfortunately, anywhere in India, as of old, a weaver's professional skill does not rate highly and accord much status, yet the work is essential to India's most important industry.

12. The Australian Connection

From the earliest days of the Colony there was a strong link with India. Officers and troops were moved between the countries, civil officials were transferred, and this close connection continued for many years. Caroline Chisholm, the outstanding early social worker who founded the Female Immigrants' Home in Sydney, came to the Colony from Madras in 1838 with her husband, Captain Archibald Chisholm, who was an officer in the East India Company forces. He was on sick leave but when he returned to India she stayed on continuing to help women immigrants and their children, and when she was later rejoined by her husband they moved between England and Australia for nearly 20 years carrying on their charity work. It is certain that she would have brought Indian cottons with her wardrobe, and with household goods to use as curtains and coverings in her Sydney home.

India was one of the places that Governor Phillip turned to for help when, within two or three years of the arrival of the First Fleet, supplies of both food and clothing began to run out. The entire community was not only close to starvation at this time but only partially clad. Watkin Tench, a young marine officer who had arrived on the transport *Charlotte*, wrote in his *Account of the Settlement at Port Jackson at New South Wales* that by early 1790 not only was the attention of everyone regardless of rank turned on food but the distress of the lower classes for clothes was almost equal to their other wants. 'The stores had been long exhausted, and winter was at hand. Nothing more ludicrous can be conceived than the expedients of substituting, shifting and patching, which ingenuity devised, to eke out wretchedness, and preserve the remains of decency'; and in the matter of patching 'the superior dexterity of the women was particularly conspicuous.' Not only the convicts were without clothes but many soldiers were threadbare and without shoes.

Muslin Dresses

It is probable also that some of the fine clothes brought out by officers' wives and families were beginning to show signs of wear in the unexpected heat and rigours of the pioneer settlement. According to fashion historian Marion Fletcher wives brought good wardrobes and were fashionably dressed in clothes of the period such as softly gathered muslin dresses and high-crown hats, while the men's mufti outfit was jacket, waistcoat, trousers and bell topper. These were in stark contrast to the convicts' rough woollen clothes, 'slops' as they were known, ill-fitting and unsuitable for the climate. By 1792 the stores situation was desperate. Bolts of cloth had been arriving from England, one of the fabrics being black 'Russia', a type of duck, but like much of the food that also came on the ships it had mostly rotted by the time it arrived. Moreover, yardages were not much help as there was little equipment in the Colony for sewing, even needles being in short supply. Ready-made clothing was required, and quickly.

At the suggestion of London, and with the permission of the governor-general of India, which was necessary in view of the strict monopoly which the East India Company preserved on all exports from India, Governor Phillip sent a store ship, the *Atlantic*, to Calcutta to purchase 'flour and pease' and other suitable food supplies, and to buy clothing for male and female convicts. Phillip stipulated that the male clothing consist of woollen jackets, flannel drawers and worsted stockings, which showed an ignorance of the Indian textile industry. In tendering for the contract a Calcutta firm, Lambert, Ross and Co. merchants licensed by the East India Company, wrote to London stating that they could fill the clothing order 'provided the woolen requests could be dispensed with and from all accounts we can learn of the climate there does not appear to exist a necessity for them. In the meantime we have got made musters [samples] of shirts and trousers of the same size as those supplied to convicts, and have sent them to Governor Phillip together with three muster pairs of shoes and blankets' — mentioning to him the price. Attached to the letter were samples of striped gingham which could be made into trousers at 2s.6d. per pair, white dungarees (denim), Patna blanket, and cotton handkerchiefs. (These samples are still attached to the original letter preserved at the Public Records Office at Kew, London.)

The Naked Colony

Distances were so great and time taken for letters and goods to travel so long, that there were always big gaps between orders and supply, and the commissariat became quickly depleted. In 1798 Governor Hunter, who had succeeded Phillip, wrote a curious missive to the Duke of Portland, then Home Secretary in charge of Colonies: 'I have pleasure in assuring your Grace that the Colony generally speaking is in perfect health; but as I am concerned to add, intirely [sic] naked for want of a supply of slop cloathing and of bedding'. No clothing worth mentioning had been received for two years. A year later there was still no clothing; local convicts were still naked and those arriving by the latest transports were in filthy rags that could not be replaced because the stores were 'destitute of every article of cloathing'. There were great fears for the health of the Colony. Real relief came with the arrival of ships, first in 1802 from London, and from Calcutta in February 1803, both bringing good supplies of clothing. The ship from India was the *Castle of Good Hope* which, as reported in the *Sydney Gazette*, the Colony's first newspaper launched in March 1803, was, at 1000 tons, the largest ship ever to enter the port. It brought a big shipment of Indian cloth and clothing and such shipments were to become regular.

Indian cottons were to become common cloth for convicts' and workers' clothes in New South Wales for many years.

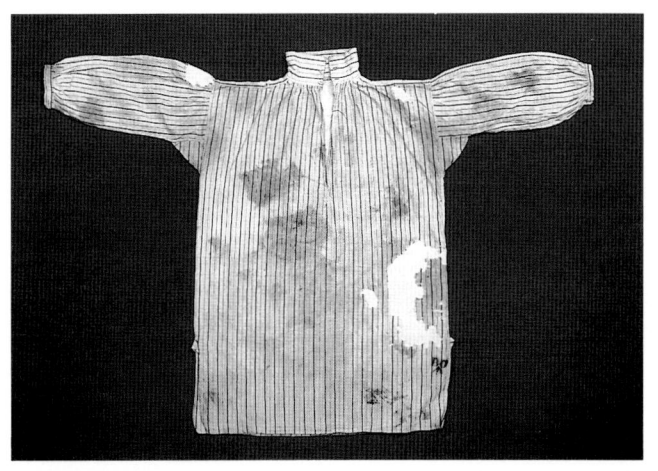

Convict shirt in blue/white striped Indian cotton discovered almost intact under the floorboards of the Hyde Park Barracks, Sydney (built 1819), during restoration work in the early 1980s. (Collection Historic Houses Trust of N.S.W.; photograph courtesy the Powerhouse Museum, Sydney)

Already blue and white striped Indian calico was standard cloth for sailors' and workers' shirts in Britain, and it was now ordered for Australian convicts. An almost intact shirt of this cloth was discovered under the floor boards of the Hyde Park Barracks in Sydney, built in 1819 to accommodate convicts, during restoration work in the early 1980s, and is now in the collection of the Historic Houses Trust of N.S.W. Made on loose lines, gathered from a high yoke with full sleeves gathered into wrist-bands, it is rather like shirts that are now fashionable for women. Dungaree, twill-woven cotton, tough and durable, in white and indigo-dyed blue, now known to the world as denim, arrived in large quantities and became the uniform of worker settlers, not only of men, who wore the overall that came to be called 'dungarees', but of women and children, to the extent that they became generally known as 'dungaree settlers'.

In Europe, Indian cottons were regarded as high quality and fashionable; in Australia, because of the 'lower class' connection they were scorned by the more genteel sections of the community. According to Paul Cunningham, a Royal Naval surgeon who spent two years in New South Wales and wrote an account of it published in London, 1828, 'Young ladies sighed after London fashions longing for China crapes and India Muslins, like the English beauties', yet, in contradiction, despised the products of the 'Eastern looms' which they considered 'too common, too cheap, too durable' — all serious defects in their eyes. At this time, wrote Cunningham, people who imported clothes from London were making fortunes, one person arriving with a stock of fashionable clothes and returning to England with £12,000, a huge sum at that time. From around 1800 others had also been cashing in on demand; for instance, officers speculated in cargoes from India which included in addition to copious amounts of spirits — rum, brandy, etc. — large quantities of cloth. Also from about that time real traders began to set up, led by Robert Campbell, a merchant from Calcutta, who had moved his business interests to Sydney after he had come to the Colony with two or three shipments of stores from India and foreseen its future.

Investment Cargoes

Advertisements for sales of ships' cargoes began to appear in the *Sydney Gazette*, with this one in its first issue for a sale of 'Investment Cargoes to be held at Mr. Campbell's on Tuesday, 8th March, 1803, of goods arrived by *Castle of Good Hope* from India.' Cargo to be sold included, in addition to live stock — cows and horses — and dry rations such as sugar, rice, pepper, etc., cotton cloth consisting of a variety of 'chintz, frocks and trowsers, soldiers' plain and frilled shirts and Blue Gurrah.' (Gurrah, which had already been used for male convicts' clothing, was a coarse Indian muslin described by one of the administrative officers as 'little better than bunting and quite unsuitable for issue as winter clothing.')

Later, in its December 1803 issue, the *Gazette* advertised a sale at the warehouse of S. Lord of 'Investments from Ship *Betsey* from Madras' including, in addition to tea and sugar, 'Madras Longcloths, Chintzes, Cambrick and muslin handerkchiefs, pumjums [mats] and gunny bags [sacks].' (Longcloth was an ordinary cotton from the Madras area and was popular because of the lengths of the pieces, about 37 yards (34 metres), whereas other cloth pieces were much shorter, around 10 to 15 yards (9–13 metres). Handloomed Indian cottons today are either in half-piece lengths, 25–28 metres, or full-piece, 50 metres. Longcloth was usually exported in white, but also came in blue or brown.) Another sale in that issue of the *Gazette* advertised dungaree at 2 shillings per yard, bandanna handkerchiefs, blue and other ginghams at 2 shillings a yard, chintzes at 5 shillings to 6 shillings a yard, and calicoes at 2 shillings a yard.

Waterloo Dresses

Advertisements after about 1815 indicate the shipments of Indian cottons were now coming via London and would have been exported through the Normal East India Company channels. Here is an interesting notice of a sale at Marr's Warehouse, Castlereagh Street, March 1816, of the latest fashions imported from London which reflected the times: 'Waterloo dresses, fashionable straw bonnets, French cambric, Scotch cambrics and shirting, English imitation shawls' and at the bottom of the list 'Indian longcloths, cossas [Bengal muslins] and calicoes.'

Investments by officers were frowned on by Governor

Advertisement in first issue of the Sydney Gazette, *5 March 1803, for sale of cargo from ship* The Castle of Good Hope, *just arrived from Calcutta — note 'Chintz' and 'Blue Gurrahs' (cheap muslin). (From copy of* Sydney Gazette, *Vol. 1, No. 1, Saturday 5 March 1803, held in Mitchell Library, State Library of New South Wales)*

Document chintz, portion of a petticoat printed by kalamkari *technique on the Coromandel Coast, India, mid-18th century, collected by G.P. Baker of the English textile printing firm G.P. & J. Baker. (Copyright © G.P. & J. Baker, Ltd)*

Modern copy by Bakers of the petticoat design on chintz called 'Curzon'. ('Curzon' copyright © G.P. & J. Baker, Ltd)

'Rhododendron Sprig', a pretty chintz design derived from a 19th century botanical print, from Mrs Monro Collection. ('Rhododendron Sprig' copyright © 1986 G.P. & J. Baker, Ltd)

'Fleur de Versailles', a 1930s adaptation of a French 18th century blockprint. (Reproduction document print from the Design Archives Collection ©)

'Raj', an Indian-influenced design first printed in Europe in the 1830s. (Reproduction document print from the Design Archives Collection ©)

'Mansfield Park', a gentle country house chintz taken from a wallpaper design of the 19th century when pansies were a popular garden flower. (Reproduction document print from the Design Archives Collection ©)

Macquarie (governed 1810–1821) who complained to London about the practice and wrote 'Officers in the Civil Departments of the Service of this Colony and Medical Officers are trafficking in various articles of Merchandise like ordinary traders', adding that such practices were clashing with their public duties and requesting instructions on how to deal with the problem. But, although Macquarie tried hard to stop the practice, it went on for some time in an underhand way.

Goods imported were becoming more luxurious as the Colony began to thrive economically as shown in this *Gazette* advertisement of 14 February 1818: 'New Investments just imported and now selling at Marr's warehouse, Sydney' . . . included 'Elegant saddles, Children's beaver hats, Ladies' fashionable shawls, White jean and Marcella [for bedspreads], Leghorn hats.' Plenty of goods were still coming from India as shown in an advertisement in December of that year: 'Sale of Investment on behalf of David Shaw' with goods including 'Bengal sugar, fine India prints and Pondicherry cloth.'

One officer who indulged most successfully in speculative investments of all kinds of imported goods was the controversial John Macarthur, a Captain in the New South Wales Corps, who later as a settler displayed extraordinary entrepreneurial skills which, backed up by the practical qualities of his wife Elizabeth, laid a solid foundation for the Colony's pastoral and productive future. Macarthur, with great foresight, secured the right to import merino sheep from Spain and was to find the first overseas markets for Australian wool. Attempts were made to weave woollen cloth from Macarthur's early clip. Unfortunately, a master weaver, Edward Wise, who had been appointed in London, drowned on the passage out and no one was found to replace him for a long time. Later, women convicts at the Parramatta Factory did produce woollen cloth for convicts' clothes.

Invoices preserved with the Macarthur Papers show that the family imported through London firms a wide range of commodities including Indian cottons, much of which would have been used for clothing the many people they employed on their farms; but as there are such large quantities there was obviously a surplus for resale. The invoices from 1812 to 1832 show several entries for Indian cloths such as mull muslin, dimity, calico, gingham, dungaree, and longcloth. There are also many cloths with names which were then common but have gone out of usage such as *baftar*, a type of calico from Gujarat, *cossar*, a good quality plain Bengal muslin, and *sannah* or *sannas*, plain cotton from Orissa which was exported to Europe in large quantities from 1640. The Macarthurs' first house at Elizabeth Farm near Parramatta, now administered by the Historic Houses Trust of New South Wales, has been restored to look as closely as possible as it did in their time, with muslin curtains, and chintzes in reproduction designs of the early to mid-19th century in the various rooms.

Apart from wool there had been various attempts to produce cloth in the Colony, one of the earliest being a coarse linen woven by women convicts from locally grown flax. This was intended for sailcloth but as the production was extremely slow, about 100 yards per month, the project was abandoned. In Governor King's time cotton-growing was tried, with seeds procured from the Bahamas, without success; but, as King wrote to London, 'There can be no doubt of it succeeding further north . . . from the latitude of 20 degrees to Cape York'; and there was a proposed exploration of the North to find suitable areas where cotton, and also indigo and maize, might be grown. By such production it was hoped not only to provide cloth for the Colony but to establish trade. As for indigo, a species of wild indigo had been found growing near Sydney, but when processed by the only convict who knew anything about it the results were unsatisfactory — 'instead of blue it gave a dirty brown' reported Governor King. A visiting French ship's officer who also understood indigo-processing got the same result, so the project was abandoned.

Australian Cotton

Cotton was later to be grown successfully in Queensland in the 1860s reaching its peak of production and sale in 1871 when, as a result of the failure of the American crops and the Civil War and subsequent shortage, any cotton that could be produced could be sold to Lancashire, and Australian growers took full advantage of this situation. Cotton in Australia was considered a secondary crop until 1961 when growing was commenced seriously in the newly irrigated area near the Namoi River in northern New South Wales. Cotton is now being grown in other districts of New South Wales and in Queensland. It is a completely modern mechanised industry producing high-quality long-staple cotton in big world demand. The basic type is *G. hirsutem*, or upland cotton, which accounts for most of the cotton grown in the world, but

Australia has developed its own *G.* 'sicala' and *G.* 'siokra'. Cotton has now become one of Australia's largest primary exports and it ranks as an important world producer. Cloth from Australian cotton is of excellent quality.

As regards early imported Indian cottons, as the years went by the sales of shipments advertised in the *Gazette* reflected the fate of the Indian textile industry and the rise and dominance of the Lancashire mills. Thus in April 1835 the word 'Manchester' appears for the first time and in due course in Australia 'Manchester' became the generic term for general household cottons. 'Manchester departments' in the big stores existed until fairly recently.

Manchester dominance, at any rate within the British Commonwealth, continued until after the Second World War when gradually other countries began to emerge as important exporters of cotton cloth, and soon there appeared in Australian shops cottons and cotton mixtures from Japan, America, and various European countries such as Switzerland, Finland, Holland and France. After Independence in India it was several years before the power mills could develop and increase production sufficiently to export in a significant way. Previously there had been limited export markets developing in surrounding countries of the Middle East and Africa, but by the late 1960s buyers from all over the world began going to India to buy cloth manufactured in the big mills of Bombay and Ahmedabad. Mill-woven cottons coming to Australia which are most obviously Indian have been calico, bleached and unbleached, and crinkle cotton for fashion clothing; but a great deal of plain cotton has been imported both for clothing and as base-cloth for the local textile industry, and there have been a variety of prints, checks and stripes, and shirting, for manufacture into everyday garments and homeware items. Mill-woven silk mainly from Bangalore is imported both for furnishing and clothing. There are also some beautiful dyed handloom wild silks on the market.

Indian Cottons Return

After the revival of the handloom and handcraft industry in the early 1960s the first goods to arrive in Australia were the block-printed Indian-style garments for which cheap unreliable dyes had been used and which gave Indian cottons a bad reputation. Next came the specialised furnishing fabrics. One of the first people to import these was Marion Best whose name is known in Sydney for having introduced, as early as the 1940s, good modern Scandinavian-type design in furniture and adventurous colour concepts in interior decorating. She travelled extensively always looking for the unusual in textiles, and introduced Australians to Marimekko cottons from Finland, Indian handloomed cottons, and Thai silks. People came to her shop from all over Australia and began to use and appreciate these unusual, artistic fabrics. When Marion Best closed her shop in the early 1970s there was a vacuum in the market for Indian handloomed cottons that Ascraft Fabrics was to fill. Marion Best's was a retail shop so the influence was correspondingly small; Ascraft wholesaled the fabrics all over Australia making Indian cottons known in every part of the country. Many of those imported were the same as the 18th century types with the same names such as gingham, birdseye, cutwork (jamdani) and mull (muslin). Others from the village looms were based on traditional weaves but had new names and were in colours and designs to appeal to modern taste, in some cases specifically Australian taste. Among textured designs with names that became well-known in Australia were Rajput, Ratnagiri, Hapur stripe, Hastina, Hissar, Sirsa, and Barsat. Jaipur plains and stripes in colours were very popular for several years. There were calico types — Goa, Bhagalpur, Goa Casement; and a favorite tabby weave, Delhi Homespun Cotton. These all sold well until the early eighties when they settled down into a smaller collection of 'classics' notably Rajput, Goa, Delhi Homespun Cotton, various types of muslins, plus some new designs in textured cottons, all in white or off-white. The checks, stripes and solid colours from southern India remain popular and are highly regarded for their good quality. Another beautiful fabric which has long been popular in Australia is Crewel, embroidered wool in white on white, or various coloured floral designs on cotton.

All the Indian hand-loomed fabrics have added their own special character to late 20th century interior decorating in Australia.

13. Epilogue – India Today

Since Independence the whole Indian textile industry, mill-woven and handloom, has been developing to the stage where it is now, near the end of the 20th century, one of the biggest in the world. Gandhi would be especially pleased with the state of the handloom sector which he himself revived more than 70 years ago with one dusty spindle and a few old wooden looms. Although no strict count has been made, with the help of government subsidies, this handloom sector has now grown to around 4–5 million looms throughout the country producing around 4,000 million metres and more of cloth per year, which represents a lot of weaving. It keeps millions of people employed, large numbers because of the Indian habit of sharing jobs, and this traditional employment for the masses was just what Gandhi wanted. A huge proportion of the handloom cloth is consumed on the local market with only about one eightieth exported but giving a very satisfactory return. The modern revival began officially in 1952 with the establishment of the government-sponsored All-India Handloom Board to co-ordinate the weaving industry under the guidance of Pupul Jayakar, eminent patron of crafts and textile scholar. She and others, like Kamaladevi Chattopadhyay, were influential in persuading the government to give the utmost assistance to craft revival generally and by their writings inspired Indian people to appreciate and take pride in their heritage.

In total production of cotton cloth, mill-woven and handloom, India ranks as a major world producer, and in exports of cotton piece goods, which include fabric metrage and clothing manufactured by all sectors, India is probably the biggest. In recent statistics India was the only country listed as having no imports of cotton piece goods. Some woven woollens are imported and some synthetic material, but India now manufactures its own synthetic yarn and the mills produce large quantities of synthetic cloth for the domestic market — easy-care nylon saris are very popular — but the proportion of mill-woven synthetic cloth to cotton is small.

Bombay Cloth Market

One of the most important textile states is Gujarat the capital of which, Ahmedabad, has been called 'The Manchester of India'. There are many large powerloom mills there and Gujarati textile representatives dominate the huge Cloth Market in Bombay called officially the Mulji Jetha Wholesale Trading Centre. This is housed in a lofty exhibition-style hall covering nearly two hectares situated not far from the Taj Mahal Hotel and other imposing British Raj buildings in downtown Bombay. Buyers from all over India and from overseas come to this market to select and order fabrics. On entering the market an extraordinary sight greets the visitor. There are hundreds of booths each belonging to a different company representing many of the one thousand or so textile mills. The floor of each booth is raised about 30 centimetres above the ground and is covered with a mattress encased in snowy white cotton, while around the sides are pillows also in spotless white cases. This 'bed' linen is changed every day throughout the whole market. Above the mattresses are shelves holding telephones, fax machines, and other modern equipment. On the mattresses recline the salesmen and buyers, Mughul fashion, making leisurely decisions from the hundreds of sample swatches on display. This style of doing business emphasises traditional practices that are an integral part of Indian life but which, though appealing, can also hinder progress in the modern sense.

Traditional methods certainly permeate the cotton-growing industry which, like India itself, is ancient, vast and complex. As a world producer of raw cotton India ranks with China, the U.S.A. and what used to be the collective U.S.S.R. Only about 7% of raw cotton is exported with the rest spun into yarn or woven into cloth for consumption on the domestic market. The quality of Indian cotton was not always highly regarded but with modern agricultural expertise growers have been able to produce a large variety classified according to staple and feel from superfine through to average. An extraordinary achievement is that India has become one of the biggest producers of extra fine percale cotton hitherto associated with Egypt. There have been criticisms of the state of Indian cotton offered for sale, one being that it is dirty and dusty which makes it difficult to spin and produces marks on the woven cloth. This occurs because of inefficient ginning and baling by the growers but the Indian government has seen the problem as urgent and set up a programme of financial aid for the purchase of up-to-date equipment. But because of the vastness of the industry improvements will take a long time to filter through.

Dye Issue

Dyes have been a contentious issue especially in the minds of Western customers. One widely held belief is that vegetable dyes are still used; another notion is that all Indian cottons fade. As to the first, vegetable dyes have not been used on a wide commercial scale for more than 30 years. In any case there are few people who can handle such a specialised method, and those few master dyers who still survive are nurtured like national treasures. Recently, however, vegetable dyeing has been returning in a specialised way and encouraged by government-sponsored organisations intent on keeping traditional crafts alive.

As for all Indian cottons fading, this is an unfair generalisation based on the fact that some of the first dyed village loom cottons that were sold in Western countries did fade because cheap dyes were used; good dyes are expensive and could not be generally afforded at that time, and it established a reputation for fading that has been hard to eradicate. There are still problems with dyes but from the mid-seventies good quality vat dyestuffs which are the most effective for dyeing cotton began to be widely used. India now has its own chemical dye manufacturing industry producing dyes for all textile sectors. The mills employ qualified dye experts and the handloom sector can receive technical advice on all aspects of weaving and dyeing from the Weavers' Service Centres set up in strategic places all over India.

The whole textile industry is greatly assisted by the research and advice of experts who are graduates of the various textile training institutes — more than 70 in various parts of India. One of the leading ones is the Indian Institute of Technology in New Delhi founded in 1960 which provides, as do many of the other schools, degree courses in textile technology and fibre sciences leading to B.Tech. and Ph.D. Courses cover all aspects of textile manufacture including design and dyeing, weaving of all types of yarns, management, marketing, product and quality control. There are also several women's polytechnics which offer courses in design, dyeing, printing, fashion designing and dressmaking. Graduates from the schools go to jobs in the various mills, or become advisers in the handloom industry.

Improvements in quality depend to a large extent on the all-over rise in the standard of living of everyone concerned — factory workers, village weavers, suppliers, sales people and the buying public. Gradually things will change and, provided political problems can be overcome, India will emerge in 50 years or so as a leading force in the world. It has certainly re-emerged as a strong textile country.

Indian people themselves have the quiet confidence of their ancient heritage. It is up to the rest of the world to acknowledge its great textile debt to India and to give long overdue recognition to the genius of the generations of craftsmen who have passed on the results of their creative talents.

Appendix 1
Fabrics for Restoration

From the early 1980s there has been a big revival of chintz in interior decorating, and the emergence of 'document' prints which are copies or adaptations of old chintz designs from the 18th and 19th centuries that can be dated to the year, or decade, when the original design was produced. They meet the modern demand for accurate restoration of furnishings in old and historic houses. Appropriate for Australia are 19th century chintzes from about 1820 on, as designs from then would probably not have reached the Colony until the 1830s or later.

Several well-established textile printing firms in England and France produce document chintzes taking the designs from samples or drawings in their own archives or from museum collections. Advances in the technology of screen-printing have made it possible to achieve accurate, high-quality reproductions; even a decade ago old chintzes could not have been copied so well. The best quality chintzes used to be hand-block printed but this method has almost died out, with only one commercial hand-block printer left in England and a couple in France. Hand-block prints are available but because of the time and patience involved in producing them and the tremendous dyeing skill needed, they are very expensive.

Two famous old English textile printing firms producing document chintz are Warners and G.P. and J. Baker. Warners were established in 1870 but can trace a textile connection back to the late 17th century to William Warner, a scarlet dyer working at Spitalfields, the London silk weaving district where French Huguenot weavers had settled. It was his direct descendant Benjamin Warner who founded the present firm which is as famous for weaving silks and velvets as for cotton printing. They have for a long time been suppliers to the British Royal Family and wove Queen Elizabeth's robes for her Coronation in 1953 as well as many decorative brocades for the Abbey. In 1970 they celebrated their centenary with an exhibition 'A Century of Warner Fabrics' at the Victoria and Albert Museum, London.

Indian Influence

Among their document fabrics are some handweaves dating from the 14th century, and a huge collection of chintzes in designs mostly from the 19th century but some from the 18th. Most are European in concept but several show strong Indian design influences, such as 'Tree of India', 1715–1725; 'Audley', with a repetitive *buta* design, 1775–1790; 'Kashmir Trellis', 1849–1860; and 'Palaquin Paisley', same period. They have also produced an interesting Classic Revival series, 1880–1910.

The firm of G.P. & J. Baker dates from the mid-19th century when George Percival Baker as a young man acted as London agent for the business run by his father George Baker in Constantinople importing English goods to Turkey and exporting Persian carpets and other Eastern goods to England. When G.P. set up his own business in 1874 it was mainly to deal in carpets; it was not until the 1890s, when his brother James joined him as partner, that the involvement in cotton printing began. Because of the firm's background there has always been an Oriental influence in Baker designs with stylised Indian or Chinese flowers prominent. Bakers were honoured in 1984 with a special exhibition at the Victoria and Albert Museum, called 'From East to West', showing textiles from the firm's archives including cloths and costumes from India and other Eastern countries collected by G.P. Baker on his travels. He took a scholar's interest in the historical aspect of textiles and wrote a definitive book *Calico Painting & Printing in the East Indies* published in 1921. A colour-plate in the book shows a brightly coloured 18th century petticoat made on the Coromandel Coast for the European market, hand-painted and resist-dyed, collected by G.P. Baker. Bakers produced a copy of the design in the fabric 'Curzon'. The bright colours were typical of original 17th and 18th century chintzes and belie the notion that old chintzes have muted colours; the 'muted' look comes from soiling and fading over years.

In addition to 'Curzon', Baker document chintzes showing Indian influences have included: 'Simla', an interpretation of a block-printed French cotton, early 18th century; 'Clive', a copy of a multi-coloured block-printed cotton, French, last quarter of the 18th century (the 18th century French textile printers either copied *indiennes* almost exactly or mixed them with French design influences such as rococo); 'Bangalore', adapted from a painted Indian cotton, early 18th century, and many others. In 1987 Bakers produced the British National Trust Country House Collection of document chintzes in designs taken from samples of old curtains and covers in various Trust houses.

Decorating Pioneer

The Trust Collection was selected by Jean Monro, daughter of the founder of the London interior decorating firm Mrs Monro. The latter was a pioneer of interior decorating in England, opening her studio in 1926. She helped the owners of many great and historic houses in the renovation and refurbishing of their homes. Her daughter carries on the firm in the same way with many of her present clients the children of her mother's clients. But she has taken her career a long way further with design commitments all around the world and was the first woman to design the interior of an ocean liner. She was also one of the first to have the idea of document chintzes, finding it more and more difficult to find old designs appropriate for the houses she was decorating, and decided to put together a collection of designs based on remnants in her own cupboards and old fabrics found in country houses. To market the collection she formed a new wholesale company, Jean Monro Designs Ltd, selling in the U.K. and abroad. Her designs reflect the 'country house' look as the descriptions quoted verbatim from her list show:

> 'Rhododendron Sprig'. Probably taken originally from a botanical print, 19th Century, but good in earlier houses particularly bedrooms as it has a lovely freshness; 'Rose and Fern'. Hand-blocked chintz. A charming all-over design of roses and ferns with white lilac on a deep blue ground. An enormous success over many years; 'Lily and Auricula'. Hand-blocked chintz. Originally traced by my mother, Mrs. Monro, from a country house she was staying in. 'Geraldine'. A design of roses and ribbons probably mid-19th Century. Ideal for bedrooms and four-poster beds.

Many of Jean Monro's chintzes are hand-blocked; in fact, she has been largely responsible for reviving this old craft in England.

Pretty English Chintzes

One of the prettiest collections of document chintzes comes from the well-known firm Colefax & Fowler which, like Jean Monro, evolved from an interior decorating business. It was started in 1934 by Sybil Colefax, a society hostess who had a flair for decorating and a need to make some money. She did the houses of her many rich friends and the firm prospered. It did even better after she was joined in 1938 by John Fowler who was described by John Cornforth in his book *Country House Taste in the Twentieth Century* as a genius, 'a master of unity and balance, of texture, colour and pattern . . .' John was a perfectionist who endeavoured to get every detail correct in style, scale and period. In his work he would pick up small scraps of worn fabric, or peeling wallpaper from old houses and keep them in his own archives. Some of these have been drawn upon for the current Colefax & Fowler classic collection. Like Jean Monro's many of the designs reflect the gentle English country house look, for example:

'Bowood' — Circa 1840–1850, English. Could be as early as 1830s. Found at Bowood House. Originally with a dot and cross ground in pink and green;
'Plumbago Bouquet' — Second half of 19th century, dated by size and position of flowers with striped ground. There was a vogue then for conservatories and greenhouses, hence plumbagos were fashionable;
'Climbing Geranium' — Circa 1830–1840 English. First produced as a block print. Original document in reds and dark greens. This was used in a small sitting room at Vaucluse House, Sydney.
'Farah' is different, it shows an Indian influence, circa 1770, probably French.

They have many more chintzes that, if not precisely date-matched, would greatly suit 19th to early 20th century houses.

Oberkampf Collection

Another English firm well-known for document chintzes is Design Archives, formerly a subsidiary of Courtaulds, best known for producing synthetic fabrics. The designs are taken directly or adapted from a huge collection of 18th and 19th century sample books which Courtaulds acquired several years ago and which were stored in a back room, almost forgotten, until on close inspection they were found to contain this potential goldmine. There were thousands — 600,000 in fact — fabric samples and design drawings, including about 40,000 from the Oberkampf factory at Jouy from 1760, only the second year of its operation, until it closed in 1843. The collection is considered so valuable that it has been treated with museum-like respect, stored in a suitably fitted, atmosphere-controlled room, with a curator in charge. The designs include:

'Appassionata' — 1870s; a combination of finely engraved roller-printed ground and hand-blocked floral spray. The passion flower theme derived from the Regency period.
'Chelsea Bouquet' — Dated 1845, this is a classic English floral chintz.
'Country Bouquet' — The background dates this design as being 1820s; it features pansies which became a popular cottage garden flower in the 18th century.
'Fleur de Versailles' — A 1930s design derived from an 18th century blockprint.
'Mansfield Park' — Another delightful chintz that features pansies and is taken from a wallpaper design of the second half of the 19th century.
'Raj' — An Indian-influenced design from the 1830s.

From Blenheim Palace

Spencer-Churchill Fabrics — many of the chintzes in this collection are copies of old fabrics at Blenheim Palace where one of the partners in the firm, Lady Henrietta Spencer-Churchill, grew up. The designs evoke the background of English country living in the 18th century. Many of their other designs are adaptations of chintz designs in the textile collection at the Victoria and Albert Museum, London.

The French Houses

Some old French textile houses are also producing document chintz with designs that are definitely different from the English ones — 'So French', one could say! Their collections are available in Australia.

Braquenié et Cie is a firm that dates back to the late 18th century. They have been makers of Aubusson carpets, tapestries and furnishing fabrics, woven and printed, and have occupied the same building in Paris since 1823. Not surprisingly they can draw on their own extensive archives for documents, as well as famous French textile museums such as Musée de l'Impression sur Étoffes, Mulhouse, and Musée d'Oberkampf at Jouy. They have many beautiful fabrics in their collection and these two are typical:

'Marquis de Seignelay' — The original was made for this nobleman's chateau by the C.P. Oberkampf factory at Jouy around 1775. The design of tropical flowers and palms was by the French artist Jean Pillement, and influenced by Indian design.
'Courteilles' — A direct copy of a late 18th century *indiennes* imported by the French East India Company, adapted in India to European taste.

Georges Le Manach is another old French house which began as silk weavers in the 1820s in Tours where they still have their factory. Their reproduction fabrics are used for refurbishing French chateaux and one special order was fabrics for the restored Orient Express. Among their document chintzes are 'Mikado' copied from an Oberkampf design, Jouy, c. 1790, showing the fashion then for 'chinoiserie'; 'Elephants', a copy of an *indiennes* imported from India, late 18th century; and a mid-19th century design 'George Sand', a reproduction of a print from the room at the chateau of Nohant where the French writer lived from 1848. Their chintzes are superbly printed on percale cotton.

Prelle is another French firm which is among the most famous of the old silk-weaving 'Maisons' of Lyon with archives extremely rich in samples and documents. The firm traces its ancestry back to 1767, to a fabric designer Pierre-Toussaint Dechazelle, with several more designers and expert weavers in the family tree, and can claim responsibility for weaving the brocade for Marie-Antoinette's bedchamber at the Palace of Versailles, and after the Revolution fabrics for Napoleon's various palaces. The firm has continued to weave fabulous brocades and in recent years has woven reproductions of their own original brocades for Versailles. Only a relatively few centimetres can be woven each week with finished cloths costing millions of francs. This century they have added high-quality chintz to their repertoire, exquisitely produced on percale cotton. One outstanding one is 'Mon Jardin', hand-screen-printed from the original design of celebrated 19th century French textile designer Jean Ulric Tournier printed originally in Mulhouse, 1850. The depth and clarity of this reproduction are astonishing.

Toiles

Toiles, one-colour prints on white cotton, were first made in the 18th century and achieved great fame as Toile de Jouy. They showed pastoral and patriotic scenes and were printed in blue from indigo, and pinks from madder which were the only available dyes. These early toiles were printed on calico imported by the English East India Company from India and indigo also came from India until the end of the 19th century when it was synthesised. The French textile printer Oberkampf, at the height of his prosperity around 1800, would travel to London to buy whole shipments of calico from the East India Company. Modern toiles are produced in 18th century colours — pinks, blues and derivations such as misty green, greys and charcoal. Many 18th century-inspired designs are available on the market, and give a charming character to period houses.

Indian Cottons

Considering that enormous quantities of Indian cottons of various types, such as calico, gingham, dimity, muslin and so on, were exported to the West from the 17th century and then copied by Lancashire throughout the 19th century, any of them would be suitable for use in restoration. Machine-made types of this kind are still in shops. And certainly any of the Indian handloomed fabrics now being imported, natural, textured ones or small blockprints, would be appropriate.

Muslin, mull or mulmul, was a most popular fabric in the 18th and 19th centuries for both clothing and furnishings, especially for window curtains and bed drapery. It came to Australia in the early days of the Colony as can be seen from the Macarthur invoices in which are listed large quantities of muslin and other muslin types: 'Cambric muslin, and Jaconet [sic] muslin'. Other Indian fabrics listed in the Macarthur invoices are ginghams, calico, cassar (high-grade muslin), Bengal handprints, blue gurrah (coarse muslin), dimities, chintz, dungaree (denim), all types which are still available and suitable for restoration.

Special fabrics are cutwork lace, or jamdani, which gives an authentic old-fashioned look to the refurnishing of an old house. Cutwork lace is now hand-woven in India on a jacquard loom. Other pretty fabrics are Festoon Muslin and Venice, both woven in seersucker technique.

Appendix 2
Glossary

ALIZARIN Red colouring ingredient found in the roots of the madder plant *(Rubia tinctorum)* and in Indian plants such as al, or aal, also known as saranguy *(Morinda citrifolia)*; and chay *(Oldenlandia umbellata)*.

ASALI TUS See Pashm.

ASHRAM A Hindu religious retreat. Gandhi's Ashram was based on discipline and vows but more influenced by politics than religion.

BAFTAR Type of calico, originally from Gujarat but later made in other parts of India; from Persian *bafta*, 'woven'.

BANDHANI Tie-dye technique resulting in spotted patterns; thus bandanna, a red and white spotted handkerchief or square scarf.

BAPU Father, affectionate name given to Gandhi.

BEETLING Beating of cloth between two pieces of wood to give smooth surface for cotton painting.

BHAVAN A shop.

BIRDSEYE Woven design of small dotted diamond, once in cotton now associated with woollen worsteds. Name derives from Hindi *bulbul chasham* — 'bulbul's eye'.

BUTA Floral motif of Persian origin closely related to the Paisley cone shape.

CALENDERING Glazing cotton by a rotary machine using heated cylinders.

CALICO Plain cotton, but in the 17th and 18th centuries in England applied generally to all cotton cloth coming from India. Gets its name from Calicut, the town on the Malabar Coast.

CARDING STRIPS Two-toothed instruments for untangling or combing yarns.

CHAPPA Wooden printing block.

CHARKHA Spinning wheel.

APPENDIX

CHHIPA Also chippa. Blockmaker, lit. carpenter.

CHINOISERIE Chinese design influences popular in Europe in late 18th century.

CHINTZ Corruption of Hindi *chint* or *chitta* meaning 'spotted', 'variegated', or 'sprinkled all over'. Modern English meaning includes plain highly glazed cotton.

CHUNDARI Rajasthan term for *bandhani*.

CINAPATTA Hindi name for silk from China.

COSSAR Also *cassa*, from Arabic-Persian *khass*, 'choice'. Good quality Bengal muslin. Name was used in England until early 19th century.

COTTON Indigenous to India. From Arabic *qutan* and Latin *cotonum*.

COUNTS Number of yarns per unit for density and weight of fabric.

CUMMERBUND From *kamarband*, Hindi/Persian word for woven waist-belt.

CUTWORK See Jamdani.

DACCA MUSLIN Muslin so diaphanous the Romans called it 'woven wind' (see Mulmul Khas). This top quality muslin was woven until early 19th century at Dacca, formerly East Bengal, now Bangladesh.

DHOBI/RANGREZ Washerman/dyer.

DHOTI Male lower garment pulled through legs.

DHURRIE Flat weave carpet in wool or cotton.

DIMITY From Greek *dismitos* meaning double-warp thread. Cotton cloth with woven ribbed pattern of stripes or checks exported in large quantities to Europe from the 17th century.

DOCUMENT Historic fabric sample or artist's design that can be closely dated, providing source for reproduction.

DUNGAREE From Hindi *dungri*, a tough twill-woven cotton cloth usually dyed with indigo and forerunner of denim. Now known as an overall.

FOULAS Indian twill-weave silk. Woven in France from late 18th century as *foulard*.

GINGHAM From Malay *ginggang* meaning 'striped' but was also woven in checks. Produced in Bengal for centuries for the Indonesian market.

GLAZING Indian chintz was given a gloss to bring up the colours. Method was starching with rice, then beetling and chanking (burnishing with a shell) until a satin-like sheen resulted. The 'glazened' finish was immediately popular in Europe.

GUNNY Sacking, woven from jute.

GURRAH A coarse muslin-type fabric of poor quality cotton.

HARAJANS Hindu Untouchables. The more dignified name was given them by Gandhi.

HIMROO Fabric of mixed cotton and silk patterned in tapestry weave to look like brocade.

IKAT From Malay-Indonesian word *mengikat* meaning to bind or tie. In India also called *bandha* from Sanskrit *bandh* to tie.

INDIENNES French term for original painted or printed cottons (chints) taken back to France by the French East India Company (La Compagnie des Indes Orientales) and still used when referring to reproduction chintzes.

JACKONET French version of Urdu *jagannath*, a fine cotton fabric heavier than muslin and later developed by the French in checks and stripes. Popular in Europe in the 18th century.

JACQUARD Loom with automatic punched card mechanism for weaving patterns, developed by Jean Marie Jacquard in 1804 and used in France and other countries from the 1820s.

JALI Lit. 'open'. Carved open-work stone screen.

JAMAWAR Clothing lengths, lit. *jama* = costume, *war* = yardage, of twill tapestry patterned woollen cloth similar to Kashmir shawls.

JAMDANI Also called 'cutwork'. Fabric woven with supplementary weft, with cut-around geometrical or sprigged motifs on a muslin base giving lace effect. Exported to England from 17th century and popular today in interior decorating.

JOUY Small town near Versailles, Jouy-en-Josas, where C.P. Oberkampf had his famous textile printing factory producing chintz and engraved copperplate prints (Toile de Jouy) from the late 18th to early 19th centuries.

JUTE Produce of Bengal, exported from 1833 to Scottish town of Dundee for sack and rope-making.

KALAMKARI Painted and resist-dyed cotton, lit. *kalam* = pen, *kari* = work.

KALANGA Also *kalga*, Kashmiri term for the Paisley cone shape.

KANI Wooden spools around which are woven various coloured wools used in weaving Kashmir shawls; also the name for the twill tapestry technique used for the weaving.

85

KHADI Handspun and handwoven cotton, peasant's cloth, adopted by Gandhi as 'Freedom Cloth'.

KHAKI Dye colour, lit. 'ashes' or 'dust' invented by a dye chemist at Commonwealth Trust factory, Calicut, in the late 19th century for camouflage in the jungle; intended for big-game hunters but adopted by the British Army.

KHARKHANA Craft workshop.

KINKHAB Indian brocade.

LONGCLOTH Plain cotton, usually white, a staple of the Coromandel trade, exported from the 17th century to Europe; the pieces were much longer than others, about 37 yards (34 metres).

MADDER Plants of the species *Rubia tinctorum*, containing red colouring ingredient alizarin, used in conjunction with mordants in Indian cotton dyeing.

MORDANT From French word meaning 'biting'. Mordants combine with plant dyes to make them fast to cotton.

MULL Also *mulmul*, Hindi for muslin, soft light fabric woven extensively in East Bengal (now Bangladesh) and exported in large quantities to Europe from the 17th century.

MULMUL KHAS Finest quality muslin woven at Dacca specially for the Mughul emperors. *Khas* is Arabic/Persian for 'select' or 'choice'. Also called *Sangati*, 'for presentation'.

MUSTER Sample.

NAINSOOK Superfine, soft Bengali cotton, from Hindi *nainsukh*, lit. 'eye's delight'. Now used for bias binding.

PAISLEY Weaving town in Scotland where so many good imitations of Kashmir shawls were made in the 19th century that the ancient cone design became known as the 'Paisley pattern'.

PALAMPORE Corruption of *palangposh*, lit. bedcover. Popular in England from 17th century as wallhangings and bedspreads.

PARAT Dye tray for blockprinting.

PASHM Kashmir shawl wool from fleece of Himalayan goats. Ultimate quality is *Asali Tus* from the goat's underbelly from which the famous Ring Shawl, or *Shah Tus* was made.

PATOLU, pl. PATOLA Double ikat silk sari length made in Gujarat, important in the Indonesian trade.

POUNCE Method of powdering charcoal through perforated outline on paper pattern for a *kalamkari*.

PYJAMA Trousers worn by Mughuls — *pai* = leg, *jama* = clothing.

RAJPUT A handloomed cotton with horizontal textured stripes now used widely for bedspreads.

SAFFRON Colour most extensively seen in India, dye is from safflower, one of the oldest dye plants.

SANNAH Also *sannas*. Plain cotton cloth woven in Orissa; went to Europe until 19th century.

SARASSA Term for chintz used throughout the East, probably from Gujarat *saras* meaning 'high quality'.

SASH From Arabic *shash*, generic term for muslin turban-cloths, white or coloured, sometimes woven with silver or gold, or coloured silks.

SATYAGRAHA 'Passive resistance', successfully used by Gandhi in his freedom struggle.

SEERSUCKER Corruption of Persian words *shir a shakkar*, lit. 'milk and sugar'.

SHAWL From Persian *shal*.

SITS Dutch term for chintz.

SLIVERS Lengths of cotton fed into the spinning wheel.

SWADESHI Gandhi's National Freedom Movement.

SWARAJ Term for 'freedom' or 'independence', much used by Gandhi.

TAFFATIE From Persian *tafta*, lit. 'glossy twist', thus taffeta. Bengal *taffaties* were popular in Europe from the 17th century.

TOILE French word for cloth, thus *toile imprimé* (block printed); *toile peinte* (painted) and *toile de Jouy* (plate print).

TOSHKHANA Textile warehouse.

TREE OF LIFE Sacred Hindu symbol of fertility and eternal after-life, dates back to ancient Babylon.

TUSSAR Wild silk from Hindi *tasar (Antheraea paphia)*, produced in Bengal.

WARP Threads stretched tightly lengthwise in the loom.

WEFT Cross threads interwoven with the warp by the bobbin.

ZARI-BROCADE Brocade woven with gold or silver threads.

Bibliography

Adburghan, Alison, *Liberty's — A Biography of a Shop*, London, 1975.

Agrawala, V.S., 'References to Textiles in Bana's Harshacharita', *Journal of Indian Textile History*, IV, Ahmedabad, Calico Museum of Textiles, 1959, 65-68.

Aiyer, H.R., 'Human Aspects in Textile Management', *The Indian Textile Journal*, Vol. 94, No. 12, 1984, 37-42.

Allen, Charles and Sharada Dwivedi, *Lives of the Indian Princes*, London, 1984.

Ansari, M.A., *European Travellers under the Mughals (1500-1627)*, Delhi, 1975.

Arasaratnam, S. 'Weavers, Merchants and Company: The Handloom Industry in South-eastern India, 1750-1790', *The Indian Economic & Social History Review*, Vol. XVII, No. 3, 1989.

Ashe, Geoffrey, *Gandhi — A Study in Revolution*, London, 1968.

Baker, George P., *Calico Painting and Printing in the East Indies in the 17th and 18th Centuries*, London, 1921.

Basham, A.L., *The Wonder that was India*, Fontana, 1971.

Bence-Jones, Mark, *The Viceroys of India*, London, 1982.

Bernier, François, *Travels in the Mogul Empire A.D. 1656-1668*, ed. A. Constable, London, 1891.

Bhattacharjee, Tara, 'Khadi — Handspun fabric', *The India Magazine*, Vol. V, No. 1, December 1985, 46-47.

Bhushan, Jamila Brij, *The Costumes and Textiles of India*, Bombay, 1958.

Birren, F., *Color: A Survey from Ancient Mysticism to Modern Sciences*, New York, 1963.

Brand, Michael, 'The City as an Artistic Center', *MARG — Akbar and Fatehpur-Sikri*, Vol. XXXVIII, 2, 93-120.

Brett, K.B., 'The Flowering Tree in Indian Chintz', *Journal of Indian Textile History*, III, Ahmedabad, Calico Museum of Textiles, 1957, 45-49.

Broudy, Eric, *The Book of Looms*, London, 1979.

Brown, Hilton, *Parrys of Madras*, Madras, 1954.

Bühler, Alfred, 'Patola Influences in Southeast Asia', *Journal of Indian Textile History*, IV, Ahmedabad, Calico Museum of Textiles, 1959, 4-26.

—— 'Indian Resist-dye Fabrics', *Treasures of Indian Textiles*, Ahmedabad, Calico Museum, pub. MARG, Bombay, 1989, 59-80.

Burnard, Joyce, 'Indian Block Printing', *Craft Australia*, 2, 1981, 18-19.

—— 'Indian Village Weaver', *The Australian Hand Weaver*, XXXI, 4, 1978, 10.

—— 'Around the World Weaving — Sheila Hicks', *Australian Home Journal*, August, 1978, 117-118.

—— 'Sheila Hicks in Australia', *Craft Australia*, Spring 1982/3, 49-56.

Burton, Anthony, *The Rise and Fall of King Cotton*, London, 1984.

Chandra, Moti, 'Indian Costumes and Textiles from the Eighth to the Twelfth Century', *Journal of Indian Textile History*, V, Ahmedabad, Calico Museum of Textiles, 1960, 1-32.

—— 'Costumes and Textiles in the Sultanate Period', *Journal of Indian Textile History*, VI, Ahmedabad, 1961, 1-52.

Chatterjee, Ashoke, 'Conserving Fatehpur-Sikri: The Designer's Role', *MARG — Akbar and Fatehpur-Sikri*, Vol. XXXVIII, 2, 121-125.

Chattopadhyay, Kamaladevi, *Handicrafts of India*, New Delhi, c. 1970.

Chopra, P.N. *Some Aspects of Society and Culture during the Mughal Age 1526-1707*, 2nd edition, Agra, 1963.

CIBA Review, 'India, its Dyers, and its Colour Symbolism', No. 2, Oct. 1937.

Claburn, Pamela, *Shawls*, Shire Publications, Bucks., U.K., 1981.

Clark, C.M.H. (Manning), *A History of Australia*, Vol. 1, Melbourne, 1962.

Clarke, Marcus, *History of Australia and the Island of Tasmania*, Melbourne, 1877.

Cornforth, John, *Country House Taste in the 20th Century*, London, 1986.

Corrie, Rebecca Wells, 'The Paisley', *The Kashmir Shawl*, Cat. Yale University Exhibition, 1975, 24-49.

Crawford, M.D.C., *The Heritage of Cotton*, New York, 1948.

Crawley, Frank, *A Documentary History of Australia — Colonial Australia 1788-1840*, Melbourne, 1980.

Cunningham, P., *Two Years in New South Wales*, Vol. 1, London, 1828.

Edwards, Joan, *Crewel Embroidery in England*, New York, 1975.

Edwards, Michael, *Everyday Life in Early India*, London, 1969.

Fletcher, Marion, *Costume and Accessories in the 18th Century*, National Gallery of Victoria, Melbourne, 1977.

Floud, Peter, *English Printed Textiles*, Cat. Victoria & Albert Museum, H.M.S.O., London, 1960, reprint, 1972.

From East to West — Textiles from G.P. & J. Baker, Exhibition Cat., V. & A. Museum, London, 1984.

Gaitonde, S.N., 'Modernisation of Textile Industry', *The Indian Textile Journal*, Vol. 94, No. 6, March 1984, 61-63.

Gandhi, M.K., *An Autobiography or The Story of My Experiments with Truth*, First published India, Vol. 1, 1927, Vol. 2, 1929, Penguin reprint, 1987.

Gandhi, M.P., *The Indian Cotton Textile Industry — its past, present and future*, Calcutta, 1930.

Gascoigne, Bamber, *The Great Moghuls*, London, 1971.

Geijer, Agnes, *A History of Textile Art*, London, 1979, reprint U.S.A. 1982.

Ghuznavi, Sayyada R., *Rangeen: Natural Dyes of Bangladesh*, Dhaka, 1987.

Gittinger, Mattiebelle, *Master Dyers to the World*, Cat. The Textile Museum, Washington, D.C., 1982.

—— 'Ingenious Techniques in Early Indian Dyed Cottons', *MARG*, XI No. 3, Bombay, 4-15.

Glazier, R., *Historic Textile Fabrics*, London, 1923.

Godden, Rumer, *Gulbadan*, London, 1980.

Gulati, A.N., *The Patolu of Gujarat*, Museums Association, India, Bombay, 1951.

Gulati, A.N. and A.J. Turner, 'A Note on the Early History of Cotton', *Indian Central Cotton Committee*, (Technological Laboratory), Bulletin No. 17, Technological Series, No. 12, Bombay, October 1928.

Hacker, Katherine F. and Turnbull, K.J., *Courtyard, Bazaar, Temple: Traditions of Textile Expression in India*. Exhibition catalogue, University of Washington, Seattle, 1982.

Handbook of Statistics on Cotton Textile Industry, 19th Edition, The Indian Cotton Mills' Federation, Bombay, 1987.

Hardingham, Martin, *The Fabric Catalog*, New York, 1978.

Harris, Alexander (An Emigrant Mechanic), *Settlers and Convicts or Recollections of Sixteen Years' Labour in the Australian Backwoods*, London, 1847, reproduction, Melbourne, 1953.

Historical Records of Australia, Series 1, Vols. I-IX, correspondence, records, and tables, 1791-1818.

Hobsbawn, E.J., *Industry and Empire*, Penguin, Reprint 1987.

Hopkirk, Peter, *Foreign Devils on the Silk Road*, London, 1980.

Indian Cotton Annual, 1984-85, No. 66, East India Cotton Association Ltd, Bombay.

Irwin, John, 'Indian Textile Trade in the Seventeenth Century: (2) Coromandel Coast', *Journal of Indian Textile History*, II, Ahmedabad, Calico Museum of Textiles, 1956, 24-42.

—— 'Indian Textile Trade in the Seventeenth Century: (3) Bengal', *Journal of Indian Textile History*, III, 1957, 59-72.

—— 'Indian Textile Trade in the Seventeenth Century: (4) Foreign Influences', *Journal of Indian Textile History*, IV, 1959, 59-64.

—— *Diary Notes on a Brief Tour of Blockprinting Centres in Western India*, February 1957, Victoria & Albert Museum, London.

—— *The Kashmir Shawl*, H.M.S.O., London, 1973.

—— 'Indian Textiles in Historical Perspective', *Textiles and Ornaments of India*, catalogue, Museum of Modern Art, New York, reprint 1972.

Irwin, John, and Katherine B. Brett, *Origins of Chintz*, London, 1970.

Irwin, John, and Margaret Hall, *Indian Painted and Printed Fabrics*, Ahmedabad, Calico Museum of Textiles, 1971.

Irwin, John, and Paul R. Schwartz, *Studies in Indo-European Textile History*, Ahmedabad, Calico Museum of Textiles, 1966.

Jayakar, Pupul, *Handloom Textiles*, pub. All-India Handloom Board, Bombay, 1973.

—— 'Indian Textiles through the Centuries', *Treasures of Indian Textiles*, Calico Museum, Ahmedabad, pub. MARG, Bombay, 1980, 59-80.

—— 'A Handloom Pilgrimage — A History of the Indian Textile', *The India Magazine*, Vol. V, No. 1, December 1985, 14-27.

—— 'Indian Fabrics in Indian Life', *Textiles and Ornaments of India*, catalogue, Museum of Modern Art, New York, reprint 1972.

Joshi, Damodar N., 'The Grandeur of Himroo and the Muslins', *The Indian Textile Journal*, Vol. 95, No. 2, November 1984, 47-50.

Journals of Indian Art and Industry — Reports written in India by British Officials and Residents; initially pub. Calcutta, 1883. Bound vols. I-XV, Nos. 1-121, dated London 1886-1913.

Khadi and Village Industries Commission Annual Report, 1957-58, Bombay.

Krishna, Vijay, 'Flowers in Indian Textile Design', Journal of Indian Textile History, VII, Ahmedabad, Calico Museum of Textiles, 1967, 1-9.

Larson, Jack-Lenor with Alfred Bühler, Bronwen and Garrett Solyon, *The Dyer's Art — Ikat, Batik, Plangi*, New York, 1976.

Leggett, William F., *The Story of Silk*, New York, 1949.

Levi-Strauss, Monique, *The Cashmere Shawl*, New York, 1986.

Mabberley, D.J., *The Plant Book*, Cambridge, 1987.

Marshall, Sir John and others — *Mohenjo-Daro and the Indus Civilization*, archaeological report on textile find, Vols II and III, Delhi, ed. 1973.

Mason, Philip, *The Men who Ruled India*, London, 1985.

The Master Weavers, Cat. Festival of India in Britain, Textile Exhibition, Royal College of Art, ed. Pria Devi, Bombay, 1982.

Mehta, Rustam J., *The Handicrafts and Industrial Arts of India*, Bombay, 1960.

—— *Masterpieces of Indian Textiles*, Bombay, 1970.

BIBLIOGRAPHY

Mehta, R.N., 'Bandhas of Orissa', *Journal of Indian Textile History*, VI, Ahmedabad, Calico Museum of Textiles, 1961, 62-69.

Mehta, Ved, *Mahatma Gandhi and His Apostles*, Penguin, 1977.

Monro, Jean, *11 Montpelier Square — Memoirs of an Interior Decorator*, London, 1988.

Monroy, Antonio, *India*, London, 1985.

Moorhouse, Geoffrey, *India Britannica*, London, 1983.

Mukherjee, Ramkrishna, *The Rise and Fall of the East India Company: A Sociological Appraisal*, Bombay, 1958.

Naqvi, Hameeda Khatoon, 'Dyeing of Cotton Goods in the Mughal Hindustan (1556-1803), *Journal of Indian Textile History VII*, Ahmedabad, Calico Museum of Textiles, 45-56.

Narayanswamy, C.K., 'Khadi in Free India', *The Indian Textile Journal Centenary Issue, 1854-1954*, Bombay, 70-80.

Natural Dyes of India, All-India Handicrafts Board, Bangalore, 1975.

Olsen, Eleanor, 'The Textiles and Costumes of India — An Historical Review', *Journal of The Newark Museum*, Vol. 17, Nos. 3 & 4, 1965, 43-49.

Panikkar, K.M., *Asia and Western Dominance*, London, 1959.

Parry, Linda, *William Morris Textiles*, London, 1983.

——— *Textiles of the Arts and Crafts Movement*, London, 1988.

The Parrys Story, A Bicentennial Publication, Madras, 1988.

Pauly, Sarah Buie, 'The Shawl: Its Context and Construction', *The Kashmir Shawl*, Cat. Yale University Exhibition, 1975, 7-22.

Peake, R.J. *Cotton*, pub. Sir Isaac Pitman, London, 1934.

Ramaswamy, A. 'Unique Silk Trading Organisation in India', *The Indian Textile Journal*, Vol. 94, No. 10, July 1984, 43-45.

Rawlinson, H.G., *The British Achievement in India*, London, 1948.

Robinson, F.P., 'The Trade of the East India Company', *The Le Bas Prize Essay*, Cambridge University Press, 1912.

Robinson, Stuart, *A History of Printed Textiles*, London, 1969.

Sahay, Sachidanand, *Indian Costume, Coiffure and Ornament*, New Delhi, 1975.

Schwartz, P.R., 'French Documents on Indian Cotton Painting: (1) The Beaulieu ms., c.1734', *Journal of Indian Textile History*, II, Ahmedabad, Calico Museum of Textiles, 1956, 5-23.

——— 'French Documents on Indian Cotton Painting: (2) New light on old material', *Journal of Indian Textile History*, III, Ahmedabad, Calico Museum of Textiles, 1957, 15-44.

Sethna, Nelly, H., *Shal — Weaves and Embroideries of Kashmir*, New Delhi, 1973.

Seymour, William, 'The Company that founded an Empire — the first 100 years in the rise to power of the East India Company', *History Today*, Vol. 19, No. 9, September 1969.

Shah, J.K., 'Textile Research in India', *The Indian Textile Journal*, Vol. 94, No. 12, September, 1984.

Spear, Percival, *A History of India*, Vol. 2, Penguin, 1965.

Storey, Joyce, *Textile Printing*, London, 1974.

——— *Dyes and Fabrics*, London, 1978.

Study of Cotton in India, Committee on Natural Resources Planning Commission, New Delhi, 1963.

Sydney Gazette, from first issue, 5 March 1803, to issue of 10 April 1841.

Taggarsi, S.R., 'Growth of Textile Training Institutions in India', *The Indian Textile Journal*, Vol. 92, No. 5, February 1982, 59-69.

Thapar, Romila, 'State Weaving Shops of the Mauryan Period', *Journal of Indian Textile History*, V, Ahmedabad, Calico Museum of Textiles, 1960, 51-59.

——— *A History of India*, Vol. 1, Penguin, 1966.

Vadgama, Usha M. and Sunanda K. Marathe, 'Ikat Saris of Orissa', *The Indian Textile Journal*, Vol. 92, No. 2, November 1981, 91-92.

Varadarajan, K.R. 'The Poor Man's Clothing', *The Indian Textile Journal*, Vol. 92, No. 9, June 1982, 91-92.

Varadarajan, Lotika, 'Kani Pashmina of Kashmir', *Indian Studies in Memory of Professor N. Ray*, Delhi, 1984, 231-240.

——— 'Homage to Kalamkari', *MARG* Publication, Bombay, 1979, 19-21.

——— *South Indian Traditions of Kalamkari*, Ahmedabad, 1982.

——— 'Textile Traditions — India and the Orient', *Shilpakar*, ed. Carman Kagal, Bombay, 1982, 24-39.

——— 'Designs on Cotton: Horizons of the Past and Present', *India International Quarterly*, 11, 4, 1984, 69-77.

——— 'Indigo — The Indian Tradition', *Laon Cei*, ed. Gert Stahl, Amsterdam, 1985.

Vijayaraje, Scindia, with Manohar Malgonkar, *Princess — The Autobiography of the Dowager Maharani of Gwalior*, Century, 1985.

Walton, P., *The Story of Textiles*, Boston, 1912.

Watson, Andrew M. 'The Rise and Spread of Old World Cotton', *Studies in Textile History*, ed. Veronika Gervers, Royal Ontario Museum, Toronto, 1977, 355-363, Note 1.

Watson, Francis, *Concise History of India*, London, 1974.

Welch, Stuart Cary, *India — Art and Culture 1300-1900*, The Metropolitan Museum of Art, New York, 1985.

Why Paisley?, Renfrew District Council Museum and Art Galleries Service, Paisley, 1985.

Wilson, Erica, *Crewel Embroidery*, New York, 1962.

Wolpert, Stanley, *A New History of India*, Oxford, 1989.

INDEX

Page numbers in *italic* type indicate illustrations.

Agra, 48–50
Ahuja, Shyam, 71
A'in-i Akbari (Book of Akbar), 50
Ajanta Caves
  ikat designs, 40
  mural paintings, 47
Akbar, Emperor, 49
  biography, A'in-i Akbari, 50
  controls trade, 49
  Fatehpur-Sikri workshops, palace, 49, *56*
  marries Hindu princess, 49
  new dye chemistry for wardrobe, 49
  talented craftsman, 49
Alexander the Great, and Greek Colony, 8
Alizarin (madder), synthesised, 21
All-India Handloom Board, 68, 79
American Civil War
  causes raw cotton shortage, 59
  influence of cotton, 7
American Cotton belt, 9
Arab traders, 9, 13–14, 44, 60
Arkwright, Richard (water frame), 11
Arthashastra, 12, 26
Arts and Crafts Movement, 21
Aryans, 7
  'Paisley' motif, 33
  respect for weavers, 11
Ascraft Fabrics, 71, 78
Atlantic, relief ship, N.S.W. Colony, 72
Aurengzab, Emperor *see* Mughuls
  Dacca muslin, 51
  himroo brocade invented for, 50
  luxury manufactures, 51
  strict Muslim, 50
  travelling tents, 49
Australia
  cotton, 77–8
  19th C. decorating, *58*, 77, 84
Avignon, Provence, 19
  French Provincial prints, 19
  textile industry, 19

Babur, Mughul Emperor, 48
Babylon
  ancient trading, 8
  'Paisley' motif, 33
Baeyer, Adolf von
  synthesised indigo, 21
Baker, G.P.
  definitive chintz book, 81
  Indian chintz petticoat, 81
  Indian petticoat & copy, *75*
Bandha, also Bandh, Bandhu, 41

Bandhani, 41
Bans, cotton
  in France and England, 19
Beaulieu, M. de
  Indian dye secrets, 20
Bengal
  Dacca muslin, 51
  plantations, indigo & jute, 60
  silk, 45
  textile state, 53
Bentinck, Lord William, 54
Bernier, Francois, 49, 51, 53
Best, Marion, 78
Birds-eye design, 42
Bissell, John, 68
Block-printing
  Indian methods, 29
  in England, 21, 30
  natural dyes, *37*
  revival of, 29–30
  traditional craftsmen & tools, 30
  workshops (kharkanas), 30
Bombay Cloth Market, 79
Brahmans
  colour symbolism, 23–4
  rise to power, 9
Braquenié et Cie
  French document chintz, *18*, 83
British Raj
  annexation of India by Crown, 59
Buddhists, 25
Buta, 32

Calico, 13, 16, 20, 73–4, 77–8, 84
  Museum of Textiles, Ahmedabad, 43
  Textile Mill, 43
Calicut
  Commonwealth Trust, 69
  oust Zamorin, 14
  Portuguese arrive, 14
Campbell, Robert
  arrival from Calcutta, 74
  Sydney trade, 74
Castle of Good Hope, relief Ship, 73–4
Chinoiserie
  chrysanthemum in, 16, 21
  influence, 16, *35*
Chintz
  calendering, 27
  French and English trade bans, 19
  glazing, 13, 27
  goes to Europe, 12
  in Australia, *58*, 74
  kalamkari secrets, 27
  palampores, 14, *35*
  Portuguese pintadoes, 14
  trade to East, 13–14

  various names, 13
Chisholm, Caroline, 72
Churchill, Winston, 62
Cinapatta, Chin-sukh, (Chinese silk), 45
Coeurdoux, Father
  letter with dye secrets, 20–1, 26–7
Colefax & Fowler
  document chintz, *58*, 82
Colour in India
  at festivals, 23
  'blue', 24
  caste colours, 23–5
  colour wheel, 23
  fading, 22
  gradations, 31
  Rajasthan villages, *17*
  recommended for community, 23
  religions & regional influences, 22
  saffron-clad pilgrims, *56*
Commonwealth Trust
  industrial problems, 61, 71
  S. India factory & history, 69
  Scandinavian designer, 71
  sell cottons to UK, Europe, 71
  Sheila Hicks, designer, 69
  workers' conditions, 71
Congress *see* Indian National
Constantinople, 45
Convict clothing, 73
  dungaree, 73
  shirt, 73, *73*
Coromandel Coast, 15–16, 27, 32, 53
Cotton
  analysis of finds, 8, 12
  boll, *10*
  cultivation in Peru, 9, 12
  decline of Indian imports, 54
  duty on Indian goods, 59
  effect on India of American Civil War, 59
  exported to Babylon, and Rome, 9
  first Indian mills, 59
  flower, *12*
  fragment finds at Mohenjo-Daro, 7–8
  from Australia, 19th C., 59
  generic type, *gossypium*, 7
  growing in Qld, 19th C., 77
  in American Civil War, 7
  Indian plant, 10
  in tombs at Turfan, 9
  inventions of Industrial Revolution, 11
  Lancashire copies, 54
  modern Australian types, 77–8
  modern Indian industry, 80

  modern types, 9
  origin of name, 9
  percallas (percale) type, 10
  role in Industrial Revolution, 7
  seen by early Greek visitors, 8
  trade bans, 19, 53
  traded by English East India Company, 7
Crompton, Samuel (Mule Spindle), 11
Cunningham, Paul, 73

Dacca muslin
  and mending, 51
  becomes heavier, 51
  description of weaving, 51
  'Mulmul Khas' Royal Muslin, 51
  poetic names, 51
da Vinci, Leonardo (Saxony Wheel), 11
Day, Francis, 54
Defoe, Daniel, 16
Delhi
  Red Fort, 50
  Shah Jahan's court, 51
  Sultanate, 51
Denim *see* Dungaree
Design
  cross-cultural, 32
  in modern India, 43
  influences, 45
  Mughul influences, 32
  National Institute of Design, 43
Design Archives
  Oberkampf documents, 76, 82
Dimity, 42, 77, 84
'Document' chintz, *75*, 81–3
Drawloom (*jhala*), 45
Dufferin, Lord, 61
Dungaree, also dungri (denim), 42, 55, 74, 77
'Dungaree settlers', 73
Dutch East India Company
  barter textiles with Indonesia, 15
  bright colours, 19
  chintz for clothing, 16, *18*
  import chintz to Holland, 15
  mourning gowns *Wentke*, *18*, 19
  oust Portuguese, 15
  regional dress, 19
  spice trade, 14
  set up factories, 15
Dyes
  alizarin (madder) and indigo, synthesised, 21
  ancient dye chemistry, 26, 31
  chlorine discovered, 21 *see also* Scheele

INDEX

finds at Mohenjo-Daro, 8
French dye chemists, 20
Indian dye secrets, 20, 27
indigo, 28
kalamkari techniques, 20, 27
khaki, 70
letter from Father Coeurdoux, 20–1
list of dye plants, 28, 31
madder, 27–8
'mauve' discovered, 21 *see also* Perkin
miracle of fast dyes, 16
modern Indian dye industry, 80
new dye colours for Mughuls, 49
revival of natural dyeing, 28
synthetic dyes, 31, 60
vat dyestuffs, 30
East India Company, *see* Dutch, English, French
Elizabeth I, Queen, 14
Elizabeth II, Queen, 81
English East India Company, 71–2, 84
bans lifted, 2
Bengal 'taffeties', 46, *46*, 54
Candle auctions, 46–7
ceased trading, 1833, 59
cotton trade bans imposed, 19, 53
decline, 47
designs to English taste, 15
East India Docks and ships, 46, 53, *53*
establish textile factories, 15
formation, 14
huge imports of chintz, calico, 16
imports muslins, 51–2
imports raw silk, 46
Kashmir shawls, 33
make political and cultural contact, 15
Mutiny, 59
plantations, indigo and jute, 60

Fabindia, New Delhi, 68
Fabrics for restoration, 81–4
Finch, William, English agent, 15
Fostat, printed cottons in tombs, 14
Free Merchants, 54
French East India Company
at Chandernagar, 16
at Pondicherry, 16
Chintz and cotton bans, 19–20
defeated by Robert Clive, 16
efficient trade, 16
import chintz and calicoes, 16
Molière's play, 16, *19*
toiles peintes and *indiennes*, 16, *18*
Frydman, Maurice, 64

Gait, Sir Edward, 65
Gandhi, Kasturbai, 63
Gandhi, M.K. 'Mahatma', 61–7
Ashram, Ahmedabad, 62–3
assassination, 67
bonfires, 64
bungalow, 63
by ship to London, 62, *63*
Constructive Programme, 64
'Gandhi' movie, 67
handloom revival, 77

hand spinning and weaving, 53
Indigo Dispute, 64–5
in South Africa, 61
inspired addresses, 64
jailed, 66
Khadi, Freedom Cloth, 62, 64, 66–7
museum, 63
Nehrus, 65–6
new compact wheel, *63*, 64
Salt March, 65
Satyagraha, passive resistance, 61
schools established, 65
Swadeshi, National Movement, 62
Swaraj, Independence, 62
textile strike and first 'Fast', 65
village tours, 64
George V, King, 62
Georges Le Manach, 83
Gingham, 42, 53, *55*, 73, 77
Gokhale, G.K., 61
Great Mughuls, *see* Mughuls
Gulati, A.N., 12
Gupta period, 320–415 A.D., 9
Gwalior, Maharani of, 65

Handlooms, revival of, 68–71
Harajans, Untouchables, 23, 62
Harappa civilisation, 7
Hargreaves, James (Spinning Jenny), 11
Herodotus, 8
Hicks, Sheila, 69–71
Bagadara, Magic Eye, *55*, 70
colour sense, 71
Kerala Collection, 69–71
Hindus
colour symbolism, 23
dress (Mughul era), 50
Huet, Jean-Baptiste, 20
Hume, A.L., 61
Hunter, Governor N.S.W., 1795–1800, 73

Ikat
double ikat/patola, 40–1
in Peru, 40
in Thailand, Japan, Borneo, 41
modern commercial yardage, *38*, 41
tie-dye technique, 40
see also Bandha
Independence
exports to America, Australia, Europe, 69
power mills, 79
revival of handlooms and textile industry, 68–71
silk weaving revival, 47
India, maps of, 25, 67
Indian National Congress, 61
Indian traders
ancient period, 8
restrictions on travel, 9
Indiennes *see* Chintz
Indigo
cloth for sale, *17*
dispute, 64–5
'English Blue', 54
exported to Europe, 29
in Australia, 77

Marco Polo sees processing, 29
plantations, 60
scale of shades, 29
synthesised, 21, 60
technical explanation, 29, 31
Indonesia, 9, 14
Industrial Revolution
affects Indian textile industry, 53
exploit workforce, 59
Lancashire millowners, 59
Paisley shawls, 32–40
spinning and weaving inventions, 11, 53
Indus Valley, 7, 26
Investment cargoes, N.S.W., 74, 77

Jacquard, Jean Marie, looms, 34, 43
Jahangir, Emperor
jewels and cloth of gold, 50
Nur Jahan, wife, 50
see also Mughuls
Jains, colour symbolism, 24
Jaipur (Amber), Princess Jodh Bai of, 49
Jamdani, cutwork lace, *3*, 52
on jacquard loom, 52
James I, King, 50
Japan
ikat *(kasuri)*, 41
trade in Indian cotton and chintz, 15
Jhala, Prince Jayasinhji, 24, *56*
Joseph's Coat of Many Colours, 13
Jouy-en-Josas
site of Oberkampf's factory, 20
Justinian, Emperor
silk monopoly, 45
Jute, plantations, 60

Kalamkari, 20–1, 27, 29, 30–1
Kamaladevi, Chattopadhay, 7, 22, 79
Kashmir shawls
East India Company imports to England, 32
embroidered shawls, 39
end of fashion, 39
fashion in Europe, 34
French designs and colours, 34, *34*
imitations Paisley and France, 33, *37*, *38*
jacquard loom, 34, 39, 43
Jewel colours, 33
modern revival of weaving, 39
Mughul tastes, 33
Napoleon takes to Europe, 33
'Ring' shawl, 33, 40
Romans buy, 33
Turkish weavers, 33
wool used, pashm, 33
Kathuria, Raj, 69
Kerala Collection
Ascraft Fabrics, 71
designs and colours, 70–1
Sheila Hicks, 69
successful sales in west, 71
Khaki, dye invented at Calicut, 70
Khadi
after Independence, 67
becomes too commercial, 66
bhavans (shops), 67
Congress uniform, Freedom Cloth, 62

high fashion, 67
'khadi villages', 64
Movement, 62
re-organisation of industry, 66
Kilburn, William, 20
King, Governor, N.S.W., 77
Kinkhabs, (brocades) and *zari*-brocades, 45, *55*
Kottahs, warehouses, 15
Kshatriyas, warrior caste, 23
Ktesis, 13

Lancashire
competition with India and tariffs, 59
dominance from 19th C. to 1940s, 78, 84
'Manchester' piece goods, 54, 60
millowners become millionaires, 59
power looms, 54
resentment by Indians, 60
'sinful imitations', 60
workers' exploitation, 59
working conditions, 59
Liberty, Arthur
Arts and Crafts Movement, 21
Indian flower designs, 40
Kerala Collection, 71
Paisley pattern, 40
Louis XV, 20
Lutyens, Sir Edwin, 50

Macarthur, Elizabeth, 77
Macarthur, John, 77
Macquarie, Governor, N.S.W., 77
Madras, English factory, 15, 54
Madder, 21, 27–8
Mahabharata, epic, 26, 44
Malabar Coast, 14, 69
Manchester, 78
see also Lancashire
Mandeville, Sir John, 10
Marco Polo
and 'chintes', 13, 27
describes indigo processing, 29
sees cotton growing, 10
Marshall, Sir John, 7
Mary II, Queen, 16
Masulipatam
best chintz, 27
dyed cottons, 13
English factory, 15
old blocks, 30
Maurya dynasty, 321–184 B.C., 11
Mill-woven cloth
Bombay Cloth Market, 79
domestic consumption, 79
early mills, 50
Mohenjo-Daro
archaeological site, 7
discovery of cotton fragments, 8
dye vats and spinning whorls, 8
Great Bath, *8*
Harappan city unearthed, 7
Monro, Jean, document chintz, 76, 82
Mordants *see* Dyes
Morris, William, 21
Mughuls, 12, 14
appreciation of crafts and arts, 49
Babur invades, 48

91

buy English goods, 14
Dacca muslin, 51
harem costumes, 48
influence on design, 32
love of flowers and gardens, 48
love of sumptuous clothes, 48
Mulmul Khas (Royal muslin), 51
new colours introduced, 49
poverty of people, 50
richness of courts, 50
silk workshops, 46
textile industry, 49
travelling tent cities, 49
Mulhouse, 20
Muslim
  colour symbolism, 25
Muslim
  Dacca muslin, 51
  18th C. fashion, 52–53
  imports to Australia, 84
  in Australia, 73–4
  Macarthur invoices, 77
  origin of name, 52
  Paisley and Lancashire imitations, 52

Oberkampf, C.P.
  buys calico, 54, 84
  factory at Jouy-en-Josas, 20
  toile de Jouy, 20

Paisley
  block-printed shawl, *38*
  embroidered shawls, 39
  fabric designs, *38*
  French term, 33
  Indian terms for design, 33
  paisley imitations, 34, *37*
  jacquard weave, 39
  origin of design, 33
  shawls, 32
  shawl fashion, 34
  weaving town, 34
Palampore, 14, *35*
Parramatta factory, 77
Pathans
  Congress Red Shirts, 64
Patolas
  by Salvi families, 40
  export to Indonesia, 41
  from Patan, 15, *38*

magic qualities, 41
pastel, 71
technique, 41
traditional colours and designs, 41
Paul, Lionel, 69
Pepys, Samuel
  'birds-eye' cotton, 42
  buys 'chint', 16
Perkin, Sir William, 21
Peru (Pre-Columbian), 9–10
Phillip, Arthur, Governor N.S.W., 72–3
Plantations, 60
Pliny the Elder, 9–10
Pompadour, Mme de, 20
Port Jackson, N.S.W.
  Castle of Good Hope, relief ship, 73
  clothes & cottons from Calcutta, 73
  clothing shortage, 72–3
  convicts' clothing, 72–3
  'investment' cargoes, 74, 77
  luxury goods arrive, 77
  'naked' colony, 73
  settlement, 72
  Sydney Gazette, 64, 73
Portuguese
  at Calicut, 12
  pintadoes, 12, 14
  seize trade, 14
Prelle of Lyon
  document chintz, 83
Pupul Jayakar
  colour wheel, 23
  India today, 79
  Mughul dye colours, 49

Ramayana, epic, 29
Revival of handlooms, 68–71, 79
  American market, 69
  entrepreneurs, 68
  S. India factory, 69
  Sheila Hicks, 69
  village weavers, 69
Roberts, Lord
  khaki uniforms for British Army, 70
Romans
  ancient trading with India, 9
  buy cottons, chintz, 9, 13
  Dacca muslin, 51
  indigo, 29
  shawls, 33

spices and silks, 45
use of monsoon, 14
Roe, Sir Thomas, 50, 52

Sarabhai, Ambalal (Calico Mills), 65
Sarabhai, Gira (Calico Museum), 43
Sash/shash, 52
Scheele, C.W., 21
Schwartz, P.R., 21, 27, 31
Shah Jahan, Mughul emperor, 50
Seersucker, 48, *55*
Silk
  Bengal 'taffetie', 46, *46*, 53, *54*
  Chinese secrets, 45
  Constantinople, 45
  disease (pebrone), 47
  from China, 39, 45
  industry, Mysore, 47
  kauseya, Indian mulberry silk, 45
  kinkhabs, 45, *55*
  modern dupion, *36*
  modern revival, 47
  raw silk exports, 46
  Silk Road, 9, 44
  sultans' workshops, 46
  tussar, 44
  Varanasi (Banaras) brocades, 47
  weavers' guild, 11
  zari-brocades, 47
Sind, Indus Valley, 7
Sits, Dutch chintz, 13
Slade, Madeleine (Mirabehn), 66
Spinning
  charkha, & foot treadle, 10
  Gandhi, 62–4
  mechanical inventions, 11
  origin, 10
  Saxony wheel, 11
  spindles at Mohenjo-Daro, 8
Stefanidas, John
  stylised Paisley design, *38*
Stein, Sir Aurel, 9
Sudras, workers' caste, 23
Surat, English factory, 15
Sydney Gazette, 63, 74

Tata family, 59
Tavernier, Jean-Baptiste, 51
Tench, Watkin, 72
Textile institutes in India, 80
Tilak, B.G., 61

Tipu Sultan, 47
Toile de Jouy, 20, 84, *see also* Oberkampf
Tree of Life design
  'chinoiserie', 21
  cross-cultural influences, 15
  on palompore, *35*
  origin, 14
  symbolism, 15

Ulysses, 45
Union Movement, 59

Vaishyas, merchant caste, 23
Varadarajan, Dr Lotika, 31
Vasco da Gama, 12, 14
Vaucluse House, *58*
Versailles, Court of, 19–20
Victoria, Queen, 39

Wardle, Sir Thomas, 47
Warners, document prints, 81
Watt, James (steam engine), 11
Weavers, Indian
  at Commonwealth Trust, 71
  guilds, 11
  independence, 15, 21
  industrial revolution, 54
  key craftsmen, 11
  low status, 12
  Mughul era, 12
  skill unrecognised, 71
  taxation, 12
Weaving
  Gandhi's looms, 62–3
  exports to West, 68–9
  handloom revival, 68–71
  high quality S. India factories, 71
  mechanical inventions, 11
  origin, 10–11
  Sultans' silk-weaving workshops, 11
  Swiss loom finds, 10
Whitney, Eli, 59
Wood, 29

Xerxes I, King of Persia, 8

Zamorin of Calicut, 14
zari-brocades, 45, 47